Skipper

Life Story of Paul H. (Skip) Duncan, Jr.
with Pearson-Duncan-Rogers Genealogical

I have mostly pleasant memories of my life and especially my parents, wives, children, siblings, grandchildren, great-grandchildren, the rest of my extended family, and those I have had the good fortune to have as friends.

Edited by Ray Rogers

Having a brother much older than oneself leads to a total lack of "knowing the details". I was fascinated by the life Skip lead, especially his childhood. The things we shared were accentuated by our twenty-two-year separation and the sharp differences in our early lives. I'll miss our long conversations about science and technology, especially the technology he developed that I used in the Navy.

Ray Rogers
April 2023

Table of Contents

Genealogical Information	7
More about My Parents, and Early Childhood: 1935 – 1943	18
Downey: 1944	26
Our Home in Bell Gardens: 1945 - 1951	29
The Desert Years and High School; 1951 - 1954	44
College Years: 1954 - 1959	54
First Marriage, Children, and Early Career: 1959 - 1970	64
Mid Career Period: 1970 - 1990	74
Second Marriage and Final Career Phase: 1990 - 1998	95
Early Retirement: 1998-2003	101
Developments in Descendants' Lives: 2003 to 2020	112
Retirement, Continued: 2003-2020	114

Preface

My seldom used middle name is Harvey, I also often drop the Jr., and only family and friends dating from high school or earlier use the nickname Skip. I wrote this because I believe it would be nice if everyone assembled something of this sort later in life, especially for the benefit of descendants. Also, the period in which I lived was interesting, and some of my circumstances and experiences were sufficiently unusual to be of at least moderate interest. I am writing this at my home in the Washington Square area of Santa Ana, California; at the time of writing I am sixty-seven years old, and it is 2003. I have attempted to make this as brief as possible and limited it to that which I recall clearly.

(*Update:* In 2020 at age 85, I made a few corrections and added a small amount of material at the end of the part of the text written in 2003 to bring the document up to date.)

Genealogical Information

This section provides limited information about ancestors, siblings, wives, children, grandchildren, nieces, nephews, aunts, uncles, cousins, and myself. Additional information about some of the people in this genealogy section, as well as information about stepchildren, spouses and others appears in the remainder of this document.

My sister, Elsie (Duncan) Kennedy, has amassed considerably more genealogical information. It is planned to organize this into a separate, more complete and detailed document at a later date.

This brief genealogical overview is divided into two sections. The first contains information about some of my mother's descendants and their families, including myself. The second section concerns ancestors, aunts, uncles and cousins. Additional information about many of those discussed in this overview is contained in the remainder of this document. Changes in the lives of descendants after 2002 are briefly summarized in the section: "Developments in Descendants' Lives 2003-2020.

Descendants of Laura Olive (Pearson, Duncan) Rogers, and Their Families

I was born on August 27, 1935, in Nederland, Colorado, a small town in the mountains above Boulder. My full name is Paul Harvey Duncan, Jr.; a nickname my family sometimes used when I was small was "Skipper", which was shortened in high school to "Skip". I am primarily of English, German, and French descent, with lesser amounts of Scottish, Irish and American Indian (I believe of the Blackfoot Nation).

I married Doris Rosina Mauro in Santa Monica, California on September 24, 1960. She was born in Rogliano, Italy, on April 18, 1937, and emigrated to the U.S. in 1945 or 1946. Our first child, Lisa Maria, was born on March 3, 1962. Next came David Paul, who was born on February 27, 1963. Our last child, Monique Laura, was born on October 6, 1964. All the children were born in Los Angeles. My nephew, Danny, Doris' sister's son, who lived with us for several years, was born on September 5, 1951. I separated from Doris in 1985, and our divorce became final in January of 1990.

Lisa has one child, my granddaughter:

Isabella (Bella) born November 6, 2004.

David has two children, my granddaughter and grandson:

Cambria born January 1, 2004 and Jacob born June 29, 2007.

Monique has one child, my granddaughter:

Jade born May 13, 2005.

I married my present wife, Stacey Claire (Emery) Bailey, on May 19, 1990, in Maui, Hawaii; it is the second marriage for both of us. Stacey was born in Westbrook, Maine on June 3, 1944. She was raised in Maine and moved to California at the time of her first marriage in 1963. She has one daughter: Bonnie Gail (Bailey) Hanson, who was born on August 23, 1964.

Bonnie has four children; my grandchildren from my marriage to Stacey: Christopher born March 18, 1987, Michelle born April 6, 1988, Michael born January 28, 1996 and Maxwell, born August 16, 2000.

Chris has one child; my great granddaughter:

Dallas, born February19, 2005.

Michelle has two children with her husband Jason Rothgeb; my great grandsons:

Skyler born August 5, 2009 and Jayden born October 23, 2010.

My sister, whose maiden name was Elsie Rae Duncan, is a year and a half younger than me; she was born in Boulder, Colorado, on March 19, 1937. She and her husband, Maurice Leroy (Morris) Kennedy, have three children: Maurice Leroy (Lee) II, born June 15, 1953, Daniel Max (Danny), born December 24, 1954, and Todd Duncan Kennedy, born February 12, 1965.

Lee and wife Lynn have a daughter: Allison Kennedy MacDonald, born May 28, 1981.

Allison has two children with her husband Shaun:

Liam, born August 22, 2012, and Riley,born July 5, 2016.

Danny has three daughters:

Amber (Kennedy) Adams born December 18, 1973 with his first wife Lori, and two daughters with his second wife, Terry: these daughters are named Tori Kennedy Baker born December 14, 1988 and Tatum (now deceased) born October 5, 1990.

Amber has two children with her husband Alan Adams: Jason born July 8, 2000 and Brandon born May 29, 2004.

Tori has two children:

Jade Kennedy born March 31, 2014, and with husband Jason: Ivy Jean Baker born February 7, 2020.

Todd has two children: Douglas born June 19,1985 with his first wife, Kimi, and Briana

Kennedy Romero born June 21, 1990 with his second wife, Dolly.

Douglas (Doug) has three children and a stepson:

his stepson Cain was born March 6, 2004; his oldest daughter, Kaeli was born October 11, 2007, next came Laila born November 6, 2009, and finally Gia, born July 18, 2012.

Briana has three children:

Ty Kennedy born November 10, 2010, Isla Kelly born February 9, 2013, and, with husband Alex: Ella Romero, born October 14, 2016.

My brother, David Raymond, is five years younger than me; he was born in Los Angeles on November 5, 1940. (A second brother from my mother's first marriage, Richard Melvin, was born February 2. 1951; he died a few weeks later from Rh hemolytic disease.) David has two children by his first wife, Darla: Diane Denise, born July 30, 1964, and Debra Rae (Debbie), born October 13, 1966. The two children by his second wife, Sharle, are: Lacey Marie born July 27, 1981, and Raymond Dean (Ray), now deceased, born April 26, 1983.

Diane, now deceased, had three children:

Maxx Weston, born May 6, 1988, Stormee Weston, born March 15, 1991, and Jonathan Rossback Jr., born July 8, 1996. Maxx and wife Kristina have two children:

Mila Alexandra Rose born December 6, 2018 and Zoey Alena Joy born June 12, 2020.

Stormee and husband Michael Ludwig have three children:

Blaze born December 2, 2013, Daija Diane born January 4, 2015, and Weston Lea River born July 29, 2018.

Debbie has two children:

Ryan Purcell, born October 7, 1986, and Jamie Archuleta, born August 30, 1994.

Ryan and wife Reanna have twins:

Mason Robert Paul and Marilyn Cassandra born November 13, 2018

Lacy has one son with her husband, Michael:

Franklin Duncan Rice, born January 15, 2015.

My mother divorced my father in 1952, and married Lorin Rogers in 1955. She had two children; (my half-brothers), with Lorin: Ray (legally changed from Ralph) Almon, born March 5, 1957, in Auburn, California, and Mark Lorin, born June 4, 1959, in Roseville, California.

Ray has one child by his third wife, Akemi:

Greg Almon, born February 7, 1991.

Akemi also had a child from a previous marriage:

Ray's stepson Bret Davis born October 2, 1975. Bret and his wife Rechelle have three children:

Morgan born April 22, 2005, Colette born May 4, 2009 and Logan born May 10, 2014.

Mark has two children by his wife, Shalois:

Scott Lorin, born October 2, 1982, and Keith Almon, born July 21, 1985.

Scott and his wife, Bree, have two children:

Rylee, born January 23, 2018 and Tyler born February 23, 2023

Keith and his wife, Jaimie, do not have children.

Ancestors, Aunts, Uncles and Cousins

My mother's maiden name was Laura Olive Pearson; she was twenty-one at the time of my birth. She married my father on February 3, 1935. She was an intelligent, attractive, 5-foot 3 inch, fair-complexioned brunette of English, Scottish-Irish, French and (1/8?) Indian descent. She graduated from a Catholic boarding school and I believe that she may have been attending college before the marriage. She met my dad at a dance at the boarding school; my aunt Dixie (Duncan) Andrew was also attending the same school and had invited my dad to the dance. Mom was born in Evergreen, Colorado, on May 3, 1914, and died at the age of seventy from cardiac complications resulting from asthma and emphysema. She smoked for most of her adult life and was overweight in her later years.

My mother's father, Ralph Raymond (Ray) Pearson, my maternal grandfather, was a salesman and entrepreneur of English and Irish descent. He had one sister, Plooma, and one brother, Benjamin (Ben). Although Ray had little formal education, he had a formidable knowledge of history and once wrote a humorous booklet about mining entitled "the Philosophy of a Mucker". He was born in Evergreen, Colorado in 1887, and died at age eighty-five or eighty-six from health problems resulting from a hip fracture. His father, Thomas Albert Pearson, my maternal great-grandfather, emigrated from Yorkshire County, England in the mid-nineteenth century. Initially settling in Iowa, he was a "freighter" who led a train of oxen teams to Denver, Colorado. He homesteaded in Evergreen, Colorado and eventually owned a large part of this area, which was later to become an affluent resort site. The extent of the Pearson Family's holdings in this area gradually diminished over time to a single parcel of about an acre. My grandfather's mother, Katherine (Powers) Pearson, my maternal great-grandmother, was born in Manhattan, Kansas, in 1855. She was Tom Pearson's second wife. Her people were "Maine Yankees", probably of Irish descent.

My mother's mother, Emma Theresa (Cone, Merrill) Pearson, my maternal grandmother, was of Scottish-Irish, French, and American Indian descent. She had one sister, Christine. Emma's father provided for her education, so although she was orphaned at about age thirteen, she was able to train as a teacher. She first met my grandfather when she arrived in Morrison, Colorado, on the train from Denver, the last leg of a journey from Boston. He drove her and another teacher in a stagecoach from Morrison to Evergreen. It was not long before they were married; it was his only marriage, and her second. Emma was born in 1886 in Springfield, Massachusetts. She died at age seventy-five from a heart attack. Her father, Cyrus Byron Cone, my other maternal great-grandfather (my mother's grandfathers were Thomas Pearson and Cyrus Cone), was of Scottish-Irish descent. Born in 1837 in Dalton, Mass., he survived severe wounds in the Civil War. He died in 1896. Emma's mother, Nellie Mae Duneen, my other maternal great- grandmother (my mother's grandmothers were Emma Pearson and Nellie Duneen), was of French and Indian descent. Nellie was born in Lebanon, New Hampshire in 1859, and died in 1899.

My mother's sister, Gertrude Pearson, my maternal aunt, was two years older than my mom. Gertrude was a pioneer in the emerging field of x-rays, completing her career as a college professor; she is listed in the 17th edition of "Who's Who in California." The Pearson land in Evergreen was passed down to her, and she eventually passed this on to my siblings and me. She only married once, a few years before her death. Gertrude died in her early eighties from heart failure resulting from Multiple Sclerosis.

My mother's brother, Raymond Pearson, my maternal uncle, was twelve years younger than my mom. After receiving his BA in theater arts at USC, he worked his way up to the position of director at a local TV station (Channel 11). His untimely death at age thirty-six resulted from a heart attack; he had suffered from bronchial problems as a child and was a very heavy smoker. Because Grandfather Pearson's siblings had no male children, the Pearson name as related to this family disappeared with the death of Raymond.

My father, Paul Harvey Duncan, was twenty years old at the time of my birth. He was born In Denver, Colorado on April 8, 1915, and died from a heart attack at the age of seventy. He smoked and drank to excess throughout much of his life. He was of German, French and some Scottish descent. When he married my mom, he was a handsome, tall, dark-skinned young man who worked as a house painter and laborer. (My complexion is midway between those of my parents.)

My dad's father, David Wesley Duncan, my paternal grandfather, was a farmer of French and apparently some Scottish descent; he died of a heart attack in his early sixties. (The presence of Scottish is deduced from the name "Duncan"; the amount is unknown, but because it is not specifically mentioned it is assumed that his lineage was less than half Scottish.) His father, James Harvey Duncan, my paternal great-grandfather, raised thoroughbred horses in Illinois; he died in 1909. His mother, Margaret Melissa (Maiden name unknown), my paternal great-grandmother, was of French descent; she died in 1900.

My father's mother, Elsie Margaret (Geipelt), my paternal grandmother, was of German descent. She was born in Berlin, Germany in 1883, and was brought to the U.S. about a year later. She died in Los Angeles in 1940. Elsie had been a high-wire performer, bareback rider and snake-charmer in the circus when she met my grandfather. She deserted him when my father was five years old. She had brothers who joined her in her circus act, but I have no additional information about them, and I do not know if she had any other siblings. I have no information about her parents.

My dad had two brothers and one sister, all of them considerably older. Albert, the oldest, was born in 1905; he left home and disappeared at about age sixteen, to reappear forty years later as a successful business owner in Clearwater, California. Attempts to locate him had failed because he had changed his last name to Harris -- something to do with railroad strikes. Albert married and he had at least one child, but that is the extent of my knowledge about him. The only sister, Nellie Blanche (she used the name Blanche), my paternal aunt, was born in 1906, and died in her early eighties from complications resulting from emphysema; she had always smoked to excess. She was married twice; I believe her first husband's surname was Thompson. I always knew her by the surname Masingale from her second marriage. Blanche had one son, Frank Marshal Thompson (name not verified) who was about twelve years my senior; a career criminal who once made the FBI's ten most wanted list, he died of a heart attack in his late fifties. David, whose wit was legendary, was born in 1907; he became a wealthy musician with his own band. I believe David had only one son, Donald, about ten years older than me. Both brothers died in their seventies, I believe from heart attacks. At well over six feet tall, they gave my six feet tall (as normally dressed) dad the nickname of "shrimp".

The only other relatives I had contact with who are related to my father are: Dixie (Duncan) Andrew, one of my grandfather Duncan's brother James' children, and her two oldest children, Jimmy and Melva. I was about a year younger than Jimmy, and Melva is about a year younger than Elsie. Dixie had one other child, Bobby, about twelve years younger than me. Aunt Dixie and Jimmy are deceased. I also have many other relatives in the Duncan family that I have never met. In addition to James, my grandfather had three more brothers: Tom, Roy, and Millard, and one sister: Chris. Tom was killed in a plane crash while herding horses with a small airplane on his large ranch near the southern border of Wyoming. (Using the noise of low-flying aircraft to control the direction of animals is useful in remote areas. Helicopters are better suited to this than are fixed wing aircraft, and drones are now sometimes used.) Chris was a famous trick rider. I heard my parents mention Tom's wife Tillie now and then but have no more information about the family at this time. One cousin's wife has apparently researched this branch of the Duncan family, with the conclusion that it is traceable to Duncan, the King of Scots; I have not reviewed the rationale, so this bit of information is included without comment.

As noted earlier, my mother married Lorin Almon Rogers in 1955. He had a daughter, Ellen, from a previous marriage. Lorin was born in North Dakota on July 22, 1904. His father, Almon Howard Rogers was from Nebraska; his mother, Cordelia Case was from North Dakota. Lorin was a stable, non-drinking and hard-working entrepreneur. When he met my mother, he had a small business near Sacramento. Sometime after their marriage, they established a restaurant in Loomis, California, about 15-20 miles north of Sacramento. This was before the interstate highway that bypassed Loomis was completed, and business was good. Later, he and my mother had a donut shop in Roseville, California, where they (ironically) also sold health food supplements. Lorin passed away in 1968.

More about My Parents, and Early Childhood: 1935 – 1943

My maternal grandmother, Emma Pearson, was a schoolteacher; her husband, Ray Pearson, was a successful salesman and entrepreneur. With their combined income they provided well for the family until the stock market crash of 1929. Although the environment they provided for my mother was culturally enriched and upscale, the focus of their attention was primarily upon my mother's sister, Gertrude, who was two years older. Shortly after the crash of the stock market, my mother was enrolled in a private parochial school in Denver, where she worked in the kitchen to pay for her tuition; she was fourteen at the time of her enrollment. A third child, Raymond, was born at about the time of my mother's enrollment in this school. Raymond shared part of the limelight with Gertrude, further solidifying my mother's position as the upstaged child. Although her position in the family left something to be desired, her relations with her parents and siblings remained strong throughout her life.

My father was raised on a farm. His father and his father's brothers were a fun-loving group whose forefathers were primarily involved with raising horses. Grandfather Duncan was something of an entertainer who played the "fiddle" at dances, and he also loved to dance. His wife deserted him when my dad was only five years old. Apparently, he was so busy running the farm after she left that my dad had little parental guidance, and he quit school after the eighth grade to help his father with the farm. By the time that he met my mother, his reputation as a drinker and lady's man were well-established.

It is easy to see why my mother fell in love with my dad; he was handsome, witty and intelligent. If she sensed impending problems based on rumors about his reputation, she probably assumed, as women have always assumed, that she could change him. I do not know exactly when this love affair began, but they were married early in 1935 after it was discovered that my mother was pregnant with me. She was twenty and he was nineteen. They settled in Boulder and my father earned a living by cutting and selling firewood; this was during the great depression and it was the only work he could find. My mother was with him in the mountains above Boulder on the day in the late summer of 1935 when she went into labor, forcing them to stop in the little town of Nederland. My birth took place there, in the tiny home of Dr. Carrie Anderson. My sister, Elsie, was born in Boulder about a year and a half later, in the spring of 1937. From the outset, she was a cheerful extrovert who gave our parents little trouble.

I took my time beginning to talk, but apparently began with reasonably clear pronunciation. Elsie evidently also had good diction. Our speech habits likely resulted from listening to my mother's reading aloud to us, as well as the good habits of speech of our parents. (They both read a great deal.) To the best of my knowledge these early years in Colorado were happy for my parents, but it was nearly impossible to make a decent living there.

When I was three years old, our parents joined my Aunt Blanche and her husband Uncle Blackie (Melvin) Masingale, and we all moved to California. The first place my parents rented in California was a tiny house situated on a hill bordering Soto Street in the Boyle Heights section of Los Angeles. It was a good time to be a child; I had abundant freedom to play and discover. It was during this time that my propensity to think in the sense of figuring things out emerged; I was always busy building complex creations that I described as "innies". I knew exactly what they were, and became very irritated when adults didn't know what an innie was. I also thought that curvy pretzels should be called criminals and a harmonica should be a pipe-a-chure; this still seems logical to me in the sense of the relationship of phonetics to objects.

At about the time I was to begin kindergarten I developed severe whooping cough. This illness and others that arose when I started kindergarten eventually resulted in my skipping this initial educational phase. I had more than my share of childhood illnesses and broken bones but was otherwise healthy. (I have always had a relatively large chest, and during the entire time period before adolescence my weight was always a little above average for my height.) I also nearly cut the end off one of my fingers when we lived on Soto St, and since it was never stitched up, it did not heal properly; it does not look bad, but the slightest contact with the side results in a sensation like hitting the "funny" bone. In those days, young children often went barefoot most of the summer, resulting in inevitable cuts; even though some of these foot and other cuts were quite serious, many parents at that time did not take children to the doctor for such things. I suffered many cuts while growing up but was never taken to a doctor for them and never received any stitches. The only immediate remedy was that most hideous potion of torture: iodine.

My brother, David, was born in 1940; I was five years old at the time. Like Elsie, he was a cheerful kid. He developed a love of playing with water, and later, fire, which got him into some trouble, but his winning ways kept our parents' wrath to a minimum.

Soon after David's birth we rented a small house next to my Aunt Blanche's home in Bell Gardens. At that time, the population of the area consisted primarily of families that had migrated from the dust bowl region, and there were few minorities. I believe it was not long after the move that my mother began to suspect that my dad might never be a good husband and provider. He was always charming and humorous, but he was also often out of work, drunk, or absent from the home. My recollection of that time was that she was often angry and probably depressed. I suspect she felt that, as in the case of her favored older sister, she was taking a back seat to the whims of another. The portion of this anger that found its way in my direction throughout my childhood likely resulted from the fact that I was probably argumentative, beginning at a young age. Other possible reasons are more speculative; under the circumstances, for example, it would be surprising if the thought had not occurred to her that she might not be in this situation if she had not become pregnant with me. I may have also represented my absentee father. Although I recall that she had some brighter times, more than a dozen years would elapse before I thought that she had reached a state of lasting contentment.

On Sundays we regularly had a fried chicken dinner next door at my aunt and uncle's nice home, where they raised chickens at the rear of their large lot. (My aunt was a wiry, nervous, pathological worker, so the home was always immaculate, and the yard perfectly groomed.) Elsie and I were in charge of doing the dishes, even though we had to stand on chairs to reach the sink and counter. We also often had the job of plucking the feathers from the chicken. At one memorable Sunday dinner, I unconsciously said something disparaging to an older guest; retribution was immediate. My aunt, who could hit harder than my mom, struck me so hard that I found myself on the floor, bleeding from a split lip. This was unusual because she and I normally got along. By contrast, my Uncle Blackie apparently did not seem to like me; this resulted in some odd incidents, like frightening me by pretending to drop me from a bridge. Like many people in that era, he was a bigot; he called my dad a n===== lover and, during a period when I lived with him and my aunt, he told me that if I ever lost a fight with a n=====, "don't come home." He was, however, intelligent, as attested by his considerable skill at building. I suspect that my interest in reading and ability to participate in adult conversation annoyed him; it is also possible that he was jealous of my aunt's affection for me. To his credit, however, he treated my sister and brother well, and never lost control of his temper when punishing me, so belt or stick was limited to my behind (or legs, when I faced him). I am also indebted to him for introducing me to the love of shooting and fishing. His feelings about me improved after I grew up, and he eventually outgrew most of his prejudices and became a tolerant and reasonably nice person.

Sometime after we moved in, my Uncle David Duncan came to California to visit my dad and aunt, bringing his son

Donald with him. Donald was about ten years my senior. This kid was apparently a genius, but an absolute monster. While he was repairing our old radio, he asked me to hold a wire, with obvious consequences. The next day we were exploring a nearby farm and climbed to the top of a stack of bales of hay, where he convinced me to take a "hay-ride". He had me perch atop a bale of hay, which weighs ninety-seven pounds, and hold on to the wire used to bind the bale together while he pushed me off. I think it was a nearly ten-foot drop. The bale landed on top of me and I couldn't stand up straight for several days.

Uncle David parked his car in my aunt's garage, and when he backed out one day, he failed to see me in the driveway. In the resulting accident I suffered grave fractures of the illium and femur in the area of the hip joint. I was in the hospital for six weeks and spent six months in a cast from the armpits down. I made the most of it, pulling myself around with my arms and playing the alligator in games with other children. This situation forced me to miss attending school during the first grade, but a tutor was sent to my home, allowing me to keep pace academically.

Soon after the cast was removed, Aunt Blanche and Uncle Blackie took me on a trip to Colorado. I was nearly seven years old. These were the days before the big interstates, and travel was an adventure. The scenery was more visible because the roads were narrower and more tortuous, so people had to drive slower. The routes taken were also more scenic. The entire approach to travel (barring emergency) was different than it is today. The objective was to enjoy the trip, not rush to the destination. People on the road acknowledged one another as partners in one of life's best experiences. Whenever you wished, you simply pulled off the road to prepare meals or sleep overnight. My love of the outdoors began during this trip. We had slept off the road somewhere in Arizona, and the next morning before breakfast I climbed to the top of a nearby hill. The spellbinding affect of early morning in the desert was soon complemented by the smell of the coffee my aunt was brewing over the campfire. The fire cast a warm glow upon the nearby ground, adding color to the gray early dawn light. When you are seven and experiencing this for the first time, it is magical.

I attended the second grade at a school approximately a mile from our home; I walked there to regain strength in my legs. I experienced my first crush at this school; she was a cute little fair- complexioned blonde. By this time, I had developed an ability that both helped my parents and provided amusement for both them and the local grocer. I did all the routine grocery shopping; provided only with a long verbal list (I am sure that I could not remember this many items now), I would set out to the market, which was about one-fourth to one-half mile away. As I loaded the items into my wagon, I mentally calculated the total. Unknown to me, my parents had told the grocer to try to cheat me out of a few pennies. I would always catch him and demand that he correct the total. He got a kick out of this but feigned anger, dragging out the ruse for the benefit of any customers who were also in on the joke.

Our parents rented a place closer to school during part of the time that I spent in the third grade. I first remember listening to the radio with more than casual interest when we lived there. Important news about the war (World War II) was often transmitted from distant lands via short-wave radio; the background static and ebb and flow of sound produced an ethereal, captivating tone, conveying the sense of great distance and importance as nothing seems to today. Both parents were working then, so Elsie and I would earn a quarter on Saturdays by thoroughly cleaning the house; in those days a quarter would buy a movie and an ice cream soda. My mother seemed happier when we lived there. I think this mostly resulted from some improvement in our economic situation. She also began to accompany my dad to bars, eliminating part of the reason for anger at my dad's drinking; the downside was that Elsie and I were often subjected to long waits for them in the car outside these bars, which were mostly western. This experience left me with something of a negative bias toward western bars and western music that I have tried to erase with age.

The tide of WWII began to turn in favor of the Allies at about this time, and there were air raid drills nearly every day. Homes with one or more family members in the military displayed a star in the window. The star was gold colored in the windows of homes where a family member had been killed; even as a child I felt the great sadness and loss of these people very deeply.

Downey:
1944

We rented a house in Downey, another suburb of Los Angeles, for part of the time while I was in the fourth grade. Like Bell Gardens, there were few, if any, minorities in this area at that time. A train track was located about ten feet from the back of the property and freight trains would pass a few times each day. I soon explored these tracks at considerable distance from the house, finding that they crossed over a riverbed on a bridge less than a mile away. Exploring the wooded areas next to the riverbed, my friends and I found a small area where a hobo with a hunchback deformity slept. Sometimes we would stay there until after dark talking with him. At night he would make coffee by boiling it in a rusty can over his campfire. He had one fascinatingly filthy old, cracked mug he would fill with about half coffee and half a combination of condensed milk and corn (Karo) syrup; he then passed it around to our little group. We loved this glorious syrupy concoction.

The temptation of the railroad bridge soon proved irresistible to the other kids and me; we knew the danger but had to explore it. As luck would have it, I was caught far out on the bridge one day when the train was coming. It was at least a 25-foot drop to the riverbed and too far from the bank to run there in time. Fortunately, a few small platforms jutted at intervals from the side of the bridge. One of these platforms, which were used to hold 50 (or more)-gallon barrels of water for the train, was close enough so that I was able to get there before the engine reached me. These platforms, which were at the same elevation as the tracks, were barely large enough to accommodate the barrels. Since there was not enough room for me on the side toward the train, I had to cling to the barrel on the opposite side.

I was usually healthy and strong, but I frequently suffered from excruciatingly painful earaches when we lived in Downey and for some years afterward. Medical attention was never sought for this problem. A routine school examination many years later showed that I had a perforated eardrum, the result of infection or physical trauma.

Only two memories of the school in Downey stand out. It was the only school I ever attended where we played soccer, which I enjoyed. The other memory was not so pleasant. One day when my sister and I were late and missed the school bus, we decided to just take the day off. We goofed off all day and went home at the usual time, thinking that we had gotten away with it: no such luck; the school had called to check on us. Punishment at home was not the end of the matter; the next day we had to report to the Vice-Principal's office, where I was to receive swats. Although virtually all parents used corporal punishment at that time, this was particularly frightening. Regardless of how bad your parent's whipping might be, you at least knew what to expect. To further instill fear, the paddle was huge. The swats were administered with ceremony befitting the enormous gravity of the situation but were not nearly as bad as I feared and were much less painful than what I was accustomed to. The worst part of the punishment was endured by Elsie, who was kept just outside the room so she could hear the sound of the swats.

Our stay in Downey ended badly. We rented the place from a friend of my parents named Jesse Rhodes. Jesse was a strange person from the backwoods of Arkansas. Our stay there began to take a negative turn after my dad criticized Jesse for beating his (Jesse's) dog. An ominous event occurred not long thereafter: after cutting himself, Jesse threw blood on the walls of the bathroom. Clearly, we should have left then, but we did not, and on one alcohol-fueled afternoon Jesse and my dad got into a fistfight. The violence lasted much longer than these things usually do, and both were badly battered and bloody by the time it was broken up by a neighbor. We hurriedly packed and moved out. My dad felt the contest was a draw, but Jesse thought he had lost. The neighbor who intervened, a relative of Jesse's, told my dad that he had better be on guard because Jesse might try to get even. The warning, which was ignored, proved prophetic.

Our Home in Bell Gardens: 1945 - 1951

When we left Downey, we moved back to Bell Gardens, where we stayed with my aunt and uncle until we got a place of our own. They had a beautiful home, but bedroom space was limited, so Elsie and I slept in a B17 engine container placed at the back of the house; to the best of my recollection, this consisted of a cube about six feet on a side, which was constructed of wood and a material like tar-paper. Many of the neighborhood parents worked at this time, and my aunt served as babysitter for them. Although many of these kids were at least my age, whenever she had to leave the house for an extended period, she left me in charge of this brood, which numbered, I believe, as many as ten.

World War II was still raging furiously at that time and there were frequent blackouts; I particularly remember that listening to the spooky program "The Inner Sanctum" during blackouts was more frightening than usual. Everyone listened to radio programs at night; these programs captured the imagination as no

TV program ever seems to. The disappointment often experienced when viewing something similar to one of the old radio programs after adaptation to TV or the movies is similar to the disappointment sometimes felt when a favorite book is adapted to these media. Gone is the rich dimension of imagination. The absence of TV also influenced life in other ways. Kids played more outdoors, of course, but on summer nights the adults in this pre-air conditioning era also stayed outdoors a lot, listening to the radio or visiting with friends and neighbors and creating an atmosphere of closeness that is missing in today's air-conditioned world. Kids played hide and seek or kick the can long into the night; the feel and smell of these nights often comes back to me when driving through open country on summer nights with the car windows down.

The war propaganda machine was in full force during this period. Most of the movies dealt with the war, and we could not wait to grow up and kill some Germans or "Japs", as we referred to the Japanese. We thought the Axis, especially the more exotic Japanese, were truly monstrous animals. One day I was with some other kids when we found an abandoned farmhouse. Peering through the windows, we could see that Japanese had lived there; enraged, we broke in and vandalized the contents, which mostly consisted of books written in Japanese. We also found an abandoned storage shed full of eggs; they were thoroughly rotten, with a truly horrible stench. This opportunity was too good to miss. We chose sides, armed ourselves with hundreds of these, and spent the afternoon in a battle that did not end until the last egg was thrown. I smelled so bad when I got home that my aunt would not allow me in the house until I stripped naked.

A riverbed that was usually dry was located near my aunt's home, providing boundless opportunities for exploration. The banks were heavily wooded, including an abundance of bamboo, and an orange grove was nearby. Overflow from the irrigation system used for the grove emptied into the dry riverbed forming a swimming hole where we would often go skinny-dipping. Several bridges spanned the river. One day while playing follow-the-leader, we began jumping off various structures. At what proved to be the end of the game that day, I was the leader, and the structure was one of the bridges crossing the riverbed. I hesitated for a long time before I jumped; it was a long way down, more than I had ever attempted before. When I landed it was as though the ground came up to hit me: hard. The bones below my knees became missiles that pushed my knees into my chin, driving my teeth through the end of my tongue. There was a lot of pain and blood, but the tongue is wide enough so that I did not lose the tip. It was some time before any of us jumped off anything again.

The relationship between my aunt and uncle and my siblings and me continued to be strong long after we moved to our own home, which was nearby. Sometime during this period, they also lived for a while in New Mexico, in a beautiful little trailer my uncle had built. He was a lead chef at the restaurant part of Thrifty Drug Store and was sent there to help initiate this part of a new store. During the summer that they were there, Elsie and I traveled to visit them via a Santa Fe Trailways bus. This long bus trip was a great adventure for us; we sat just behind the driver at the front of the bus and did not miss a thing. I was about ten years old, and Elsie was eight or nine. Uncle Blackie once took me hunting while we were there, and he took Elsie and me fishing at Elephant Butte several times. The fact that I enjoyed these activities seemed to be about the only thing he liked about me at that time.

My parents were both working at this time, and they bought a one-bedroom house less than half a mile from my aunt's place. We were to live there for the next five years, until shortly before my sixteenth birthday. At first, things were better than they had ever been. This changed one morning when my aunt and uncle called Elsie, David and me together to tell us that our parents had been in an accident. (We stayed with my aunt and uncle on nights when our parents were working.) Since they had not been killed, it was several days before we fully comprehended just how bad it was; they did not tell us that the doctors were not sure if my dad would live. Jesse Rhodes had waited about six months to carry out his threat to get even. At that time my parents were working at the same bar, and Jesse had hidden in the dark with a large truck tire iron, attacking them late at night after the bar had closed. He was trying to kill my dad. While trying to get between Jesse and my dad, my mom suffered a cut over the eye; when she was released from the hospital a day or two later, the eye was blood red, and she had a bandage above it. My dad came home several days later. His face and head, which had been shaved, was a mass of stitches; the sinus on one side had been shattered, the cheekbone was pushed back a good inch, and his skull had been penetrated in several places. Jesse eventually served six months for attempted murder.

Jesse's attack indirectly accelerated the final phase of the demise of my parent's marriage. About the only thing my dad took personal pride in was his appearance; the loss of his looks meant the loss of one of the few things he felt that he had to be proud of. His drinking graduated from merely very frequent to more or less continual.

The long-term effect on me was devastating. Up to this point I have described my dad's behavior with the objectivism and hindsight of maturity, only barely touching on the qualities that made me idolize him as a young child. From my viewpoint at that time, the traits that made him a poor husband were not things that were obvious to me or otherwise significantly tarnished my image of him. Had he hit my mom or something, I am sure I would have felt different, but he did not, and even when he was drunk, he was never mean. I am sure that I had some inkling of the connection between his behavior and my mom's unhappiness, but I do not remember being certain of it until I was about thirteen. In his relationship toward me, he was an extremely loving father. He loved animals and seemed nearly free of hatred and prejudice. He would sit and talk to me for hours, telling all sorts of stories, many of which were unfit for kids. He was more like a renegade uncle than a father, and we were very close. Over the next few years, however, his disintegration was apparent, even to me. He spanked me occasionally and I dreaded this because when it was necessary, he was extremely angry and hit very hard. However, he never lost control when whipping me, and limited the belt to my backside.

Under the circumstances, my parents fought more and more often, and my mother would also sometimes join my dad and drink too much. The effect that alcohol seemed to have on my mother was one of unleashing stored anger; we soon learned that it was best to stay out of her way after she had been drinking. Understandably, the disintegration of the marriage accelerated. At about this time, my mom became a Jehovah's Witness and had me believing the world would end in a fiery conflagration to take place in 1959. Soon, every unusually red or orange sunset made me wonder if the schedule for the end of the world might have been moved up and the world might be starting to catch on fire. She confided in me a great deal at the time; the overall tone was one of sadness. Her message was clear and supported by biblical quotations: life is miserable, and the only peace is to be found in the afterlife. I felt sad to the point of tears for her lot in life.

I had an abundance of worries. The marriage was degenerating, our parents often stayed out too late, the world might burn up soon, and I lived in constant fear that Jesse would come back and kill us all. As I grew into adolescence, I began to think about locating Jesse and effecting a final solution to at least this last part of my worries. Fortunately, the time devoted to these horrors was mostly confined to sleepless nights, leaving days during which I was primarily occupied with many other, more normal and enjoyable aspects of life. The imprint that Jesse Rhodes left on my psyche, however, has never been erased.

I attended the fifth and sixth grades at a small school a little more than a block from our home. The school was surrounded by eucalyptus trees and had a small gazebo at one end of the dusty playground. It was at this school that I learned how to deal with schoolyard bullies; they were typically looking for easy targets, so when I stood up to them, they sought their prey elsewhere. In one instance, I agreed to meet one of these apprentice jerks to settle our differences at a specified time well after school hours, in a field near both our homes. I got there on time and waited for about half an hour, yelling his name. He lived nearby, so I'm sure he could hear me. He never showed up, and thereafter treated me with considerable deference. Other memories associated with this school are more pleasant, such as the little old-fashioned combination grocery store and gas station that was located near the school. The gas pumps were of the old hand-pump operated type with a glass top that allowed the gasoline to be viewed. Two old men ran the store; it was dark and musty and had an old wood floor that was loose in places and moved under our feet.

We would often stop there on the way to or from school, ogling the large candy bins and occasionally buying some candy.

In that time period, a few people had yet to buy their first refrigerator. The ice companies gave these people a small square cardboard sign, near the four edges of which were printed numbers corresponding to four different quantities of ice; those needing ice would put this sign in their window and orient it so the desired quantity would be at the uppermost part of the sign. The iceman, reading the number from the street, would then use large tongs with sharp tips to carry the specified blocks of ice to them. On hot summer days while the driver was carrying a block of ice to someone's home, we would climb into the back of the truck to pilfer a shard of crystal-clear ice, which was sometimes as large as an ice cream cone. It was dark, damp and cool inside, with a musty but pleasant smell. By the time I entered the sixth grade, these trucks had become a part of the past. The Helms Bakery used a similar system. They supplied people with a square sign with a big "H" that would be placed in the window if baked goods were desired. When we lived with our aunt, she would occasionally buy cream puffs from the Helms man and share them with us.

A large field devoted to various crops was located nearby. The neighbors called it the Jap field, because Japanese had farmed it before the war. A Filipino family owned it when we lived there. In the summer when the tomato crop was in, we were allowed into this field under the conditions that we were to be careful to avoid damaging the plants and take no tomatoes from the field. We would bring a salt-shaker and eat our fill of plump, juicy beefsteak tomatoes while sitting in the shade of the rows of plants. A little farther away, a large area was devoted to storing bombers in various stages of dismantling in preparation for conversion to scrap. The basic structures of many of these were still reasonably intact. Elsie and I would often get up early and taking a thermos of hot chocolate and some sandwiches, make the short journey there. We passed a large field of damp, tall grass on the way; the sound of a meadowlark in the morning mist just as the sun began to break through added to the moment, which was as much an experience of feeling as of seeing. When we got to the bombers, we spent many hours conducting imaginary bombing raids over Nazi Germany.

Soon after we moved in, the atomic bombs were dropped on Japan (August of 1945), heralding the end of World War II. I remember picking up the paper in our front yard and running into the house to tell my parents about it. This event led me to questions about the nature of the atom, initiating interest in physical science. The war ended soon after that. My first post-war memories were of chocolate bars, spearmint gum and bubble gum, none of which were available during the war. Most of my memories of these years were pleasant. I was a typical boy of that time and place, playing outside most of the time, often covered with dust, my pockets bulging with rocks for use in an ever-present slingshot.

Our house was frequently the scene of parties. My dad, always the (usually well "lubricated") life of the party, was a reasonably good banjo player. Music was often blaring on the radio or phonograph and people often danced. With the exception of the big bands of Glen Miller et. al., the music they played on the radio was pretty corny. Our house became a sort of crash pad for various affairs. Since I was always something of an insomniac and I slept in the same room as various overnight guests, I received an early vicarious education in the ways of the world. Because things did not get interesting until late at night, Elsie and

David, who had no trouble sleeping, were likely unaware of most of what was going on. The situation grew even more convoluted when a recently divorced female friend with an insatiable sexual appetite set up housekeeping in a small house trailer in our large yard. Although the overall environment was not exactly ideal for children, it did provide laughter, partly offsetting the depressing aspects of our parents' marriage.

During this time, we would often take trips to Santa Monica where my grandmother Pearson, Aunt Gertrude and Uncle Raymond lived. Getting there and back in those pre-freeway days was always a lengthy affair. Their apartment smelled of mothballs, and we had to call our grandmother "mother" because of her sensitivity about age. The Pearsons were a well-educated, academically oriented family and books were everywhere; Uncle Raymond was the first person who was able to give me some insight into the nature of the atom. The Ocean Park Amusement Pier was visible from a balcony off a back bedroom, and we could walk to the beach. We had a wonderful time there, swimming in the ocean and going on various rides at the amusement center.

I do not think we had been in our home more than a year or two when the back of the place caught on fire. Most of the kitchen and bedroom at the back of the house were a total loss before the fire department managed to put it out. When the fire was at its peak and thick black smoke was pouring out of the front door, it was found that the cat was missing; my dad immediately crawled in on hands and knees to rescue it. Rebuilding the rear of the house seemed to take forever. It was just after Christmas; at night we huddled in the living room, the cold air streaming in from broken windows. My mom did not want the living room painted because she liked the wood paneling. The wood was smoke blackened, but she intended that the entire room should be scraped clean. The paneling was cut in an ornate style, making it virtually impossible to scrape completely clean; thereafter it always smelled of smoke and looked dark and depressing. Adding to my father's depression triggered by the house fire, his father died.

Money became scarcer than ever. When the gas company terminated our service, my dad bypassed the meter with some sort of kluge that soon leaked so badly that the neighbors, in fear of their lives after smelling the gas, called the authorities. Later, when the electric company also terminated our service, he bypassed the meter. This worked, but the system then had no fuse. One night a short developed in an iron cord, creating a shower of sparks that persisted until the cord was burned completely through. It is little wonder that I worried about my parents.

Throughout this period, my mother tried to offset the negative aspects of our environment by reading to us and encouraging us to read and ultimately make something of our lives. I particularly recall poetry by Kipling, Fields and Poe, as well as stories by Kipling. She would often speak of rising above the "common herd", a phrase likely used by her elitist mother. Employing phrases such as this when dealing with children who are trying their best to fit in with the "common herd", however, falls on deaf ears. One of the few positive examples provided by my father was the habit of reading. Although he apparently devoted little thought to his choice of books, he recommended a couple that I later found were classics.

On several occasions during this time, an uncle who lived about five miles away in Huntington Park brought groceries to us. This was our Uncle Joe Andrew, the husband of Dixie, my dad's cousin. We often visited Uncle Joe and Aunt Dixie and their two children, Jimmy and Melva. As mentioned in the Genealogy Information section, I was about a year older than Jimmy and Elsie was about a year older than Melva. At that time Huntington Park was a very nice area, much nicer than Bell Gardens. We would usually stay overnight when we visited them; the kids had a sort of sleeping porch, and we would read comic books all the next morning. Uncle Joe was very ambitious and money-oriented; he usually worked two jobs so they could have the newest and best of things. I was amazed at how much these kids got at Christmas, and equally amazed that Jimmy was not nearly as happy with this incredible array of booty as I was with a couple of inexpensive gifts. Very few people had television sets then, but they had one. It had a small screen in the shape of a circle with the top and bottom cut off. Color would not become available for many years, and the programming was incredibly boring.

I entered the seventh grade at Bell Gardens Junior High School, which was about a mile from our home, in 1947. It was not the best environment; for example, it was rumored that some kids actually smoked marijuana on the school grounds. I handled a couple of bullies in much the same way as in the primary grades, with the result that they soon left me alone.

When I was in the sixth grade, tests conducted by the school showed that I was adept at differentiating musical tones, which is an indicator of potential musical ability. Since my Grandfather Duncan had left a violin to me, I took some lessons and learned to play well enough so that I was able to join the junior high school orchestra. This association resulted in travel, sometimes to such exotic places as Fullerton, in Orange County. (I did not enjoy playing this instrument, and gave it up as soon as I left junior high school.) I also got to travel to many additional communities after I entered the eighth grade and managed to get on one of the school's football teams.

My mode of dress in the seventh grade was somewhat shabby, much as I had dressed in earlier grades. This changed dramatically in the eighth grade. Although I had always had infatuations, the explosion of my hormones and abundance of pretty girls dictated preoccupation with my appearance. Since my mom was working most of the time, I learned to iron my own shirts (wrinkle free did not exist yet) and took charge of my jeans, which had to be kept dark and unfaded as long as possible. Elsie would sometimes surprise me by ironing my shirts. She was a great helper, friend and ally, often warning me when I might be headed for trouble with my parents.

Like most boys at that time, I was fascinated with cars. (My dad's fits of temper while he worked on our various older, unreliable cars, however, left me with a less than positive attitude about spending much time at this activity. Notwithstanding this bias, I was always interested in the details of how they work, just as I was interested in how everything works.) In those days nearly all cars had manual transmissions. At first, I practiced using the clutch and shifting gears, without starting the engine. Before long I would start the engine and pull the car forward and backward in the driveway. The inevitable occurred one day at about the time I completed the eighth grade. My parents had gone somewhere and I took the car out for a drive. Things went well for about a block, until I had to turn. The turn went fine until the point at completion when the wheel had to be released: I did not. The result was that I was heading toward a building; luckily, I stopped just in time. One of my friends was riding with me; we pushed the car all the way back home. My parents did not find out until years later.

At age thirteen or fourteen I was still the recipient of corporal punishment; in fact, the primary recipient within the family. David was the cute baby always loved by everyone, but he got into enough trouble to warrant occasional punishment. I do not remember Elsie ever being punished; not only was she a girl, but to the best of my knowledge she never argued or did anything wrong. My father occasionally let me have it for serious infractions, but most of my mother's loss of control had always resulted from my penchant for argument. Much of her anger was likely my fault because I was probably aware of many of the points of view that would set her off, but expressed them anyway, stubbornly refusing to relinquish my point of view. Her attitude was sometimes confusing to me; she loved to argue and encouraged us to express our opinion, but the limitations on what it was considered acceptable to argue about were not always clear to me. At no time did I ever feel that arguing constituted what some called back talk, which I understood to be defiance of direct orders.

During this time period, I began to suffer from extremely painful toothaches. A "dentist" was finally found who would pull the problem tooth for a sufficiently small fee. I think this guy would have been a standout in a torture chamber. The extraction of the badly abscessed tooth was so painful that it caused me to lose consciousness during the procedure. Notwithstanding meticulous dental hygiene, I had periodic tooth pain until my mid-twenties, requiring considerable dental work.

The solid, positive social world I had established for myself came to a crashing end during the time that I was in the ninth grade. My father was arrested for receiving stolen goods, mostly plumbing supplies that friends had stored in a shed behind the house. The event was all over the front page of the local newspaper. One of my two best "friends" walked up to my desk and ceremoniously dropped the newspaper on the desktop, asking me what I knew about it. It was over a year before I spoke to him again, and then we exchanged only a few words. I gave up any thought of being a well-behaved, studious kid; by the time I entered the tenth grade I had started smoking, drinking, getting into trouble and socializing with less than ideal kids. At one point I even gave myself a tattoo. My parents were furious.

Toward the end of the ninth grade, Elsie and I became good friends with two daughters of a friend of our parents. Through them we met some guys who had a hot '34 Ford sedan that had been lowered in front. We rode around with them a lot, going to neat places like the Pike amusement pier in Long Beach. I also had other friends who had cars; most of these guys were much older than me. The cars were really sharp then; my favorites include the modified deuce ('32) coupes and later models with all the chrome except the chrome for the grill removed. Some drinking naturally ensued, and I also managed to get into more serious trouble at about this time, for which I was sentenced to three years of probation. Based on improved attitude and behavior, this was terminated about a year and a half later, after I had moved to the desert.

I acquired a new best friend soon after I entered the tenth grade. His name was George, and he lived with his mother and her common-law husband down by the river. His common-law stepfather looked as though he had just been released from prison; he had huge biceps and his arms were covered with tattoos. He entertained George and me with all sorts of tales of past illegal activity. One day after coming home from the river, a dried leaf from something we were carrying fell into an ashtray. After sniffing the resulting smoke, George's stepfather said he thought it was marijuana; naturally, George and I returned to the river and retrieved some more, which we dried and attempted to smoke. We tried to convince ourselves that it was affecting us, but I doubt that it was marijuana. The coalition of George and me was not a good one; we ditched class for much of the tenth grade.

Beginning with the first or second grade, I nearly always had a crush on one girl or another, but I did not have a steady girl friend until I was fifteen, when I began to go steady with a certain brunette. She wore my ring on a ribbon around her neck, and the relationship was about as serious as it gets at that age. We listened to music a lot. This was the age of the ballad, which was a considerable improvement over some of the inane songs of a few years before. In an incident associated with this relationship I had a brief physical altercation, the only one of any significance I had been involved in since the sixth grade.

My mother had another baby, Richard Melvin, on February 17, 1951. The name Richard was probably chosen from the name of one of our family's friends. Melvin, it will be recalled, was my Uncle Blackie's name. The baby died a few weeks later from Rh hemolytic disease.

One day in early June of 1951, my Grandfather Pearson came for what I thought was just another one of his infrequent visits. I didn't know it, but this visit was different. My mom had made plans to separate from my dad. We were going to move to the desert, where my grandfather ran a gasoline station.

The Desert Years and High School; 1951 - 1954

My mother, my brother and I arrived in Red Mountain, California, in June of 1951, about two months before my sixteenth birthday. (My sister stayed behind with a friend of the family. She joined us a few months later.) Red Mountain is one of four small towns constituting the Rand Mining district, which is in the Mojave Desert approximately 150 miles northeast of Los Angeles. Randsburg, about two miles northwest of Red Mountain, was the center of large mining operations in the late 1800s and early in 1900s. During the time that I lived in Red Mountain, Randsburg had a population of about a thousand. Randsburg is now nearly a ghost town with few residents, but five thousand people, mostly miners, were settled there in its heyday during the late nineteenth and early twentieth centuries. Johannesburg, a residential community, is located about two miles north of Red Mountain. Johannesburg had a population of about a thousand in the early 1950s; it presently has the largest population of the group. The fourth town, a tiny settlement named Atolia, is located about two miles south of Red Mountain. Atolia was the site of tungsten mining operations until the late 1950s. These mines provided the primary support of the residents of the other three towns during the time that I lived on the desert. A couple of families and a few miners lived in Atolia at that time, but no one lives there now.

My grandfather had two mining shacks in Red Mountain; these shacks are still there. (My sister and her husband, who use them for occasional getaways, have extensively refurbished them through the years and added another small building with a flush toilet and a shower.) The shacks are situated one behind the other, spaced about fifteen feet apart. My grandfather lived in the tiny shack nearest the highway; it is about twenty feet from Highway 395. We were to live in the larger shack, which is about twelve by fifteen feet; it was divided into two rooms: the narrow front room contained a wood stove on which my mom prepared our meals and the room at the back became our bedroom. Cardboard had been put up on the inside to provide some insulation, and a thick layer of dust covered everything. At that time the shacks had no water and the toilet consisted of an outhouse located a short distance from the rear shack. I felt that I had arrived in Hell. My mother began to cry, so I put my arm around her and told her that I would take care of her. The time of my adolescent irresponsibility was over.

I immediately went to work at my grandfather's service station. I worked all day, seven days a week, so I had little time to lament the situation. There were three bars at that time, in a town with a population of three hundred residents plus about a hundred prostitutes. (One of the bars burned to the ground within a few days of my arrival, never to be rebuilt.) There was no law enforcement, so troublemakers were simply severely beaten by the bar owners, their hired goons, or an informally appointed constable. When it was felt that the offender had been adequately punished, he was ushered out of town. Noisy revelry continued far into the night. The bar owners, who were also pimps, ran the town. In the manner of Mafia bosses, they ensured maintenance of the status quo by treating the residents of the town well. Similar to areas dominated the Mafia, some residents actually appeared to admire them, ignoring the source of their money and power. This description of these characters is based on the objectivity of hindsight; at the time they were merely one part of a very strange scene. Far more interesting were the prostitutes, many of whom were young and attractive.

About two months after arriving in Red Mountain, I was hit by a car. When the accident occurred I was returning to work from my dinner break, walking well off the side of the road. I did not hear the car approaching from behind until the instant just before impact. I was to learn later that my grandfather was driving the car; he was so drunk that he did not know he had hit me. I suffered severe fractures of the tibia and fibula and had to be kept in the hospital in traction for thirty-three days. A hole was drilled through my heel, through which was threaded a stiff wire. To keep the leg in tension, the wire was connected to a system of pulleys and weights. The pain was so bad that I was kept on morphine for ten days, after which it was withdrawn to avoid my becoming addicted. The remainder of the thirty-three days was a nightmare of pain. At the end of this period, my leg was encased in a cast. The effort of dealing with the pain resulted in a loss of nearly sixty pounds. Upon returning home I looked like a skeleton. It took me about a year to regain normal weight.

The accident and long recovery constituted a major turning point in my life. Although I later played football and engaged in heavy work in the mines, my outlook became less physical and more intellectual than might have otherwise been the case.

Although I had little interest in schoolwork, I became fascinated with the fundamental philosophical issues of mankind and read such works as the Rubaiyat of Omar Khayyam and the book of Ecclesiastes. Later, in a college course exploring literature in greater detail, I discovered the incredible depth of understanding of the human condition achieved by Shakespeare and began an objective study of the world's religions; the inevitable consequence of which was reluctance to embrace any particular organized religion. This philosophical turn of mind lasted into my early twenties, and, to a lesser extent, for the rest of my life.

Soon after I returned from the hospital, we moved to a small house located near the gasoline station where I worked. Elsie joined us at about this time, and we soon attended the Rand District High School, which was located in Randsburg. A school bus picked up the kids living in Atolia, Red Mountain, Johannesburg and a small settlement some distance from the mining district. Because I had ditched most of the tenth grade, I had to repeat it. Randsburg High School included grades nine through twelve. Although some of the faculty members were very competent, the average quality of the classes there was relatively poor. The social atmosphere was, however, very good, and I quickly made friends. The kids there were much more clean-cut and normal than my recent friends in Bell Gardens. The guys wore clean jeans or khakis, shirts buttoned to the neck, and well-shined shoes similar to the military type, but with thicker soles. The girls wore sweaters, scarves, hoop skirts, and ballerina shoes or saddle shoes and bobby socks. Nearly everyone smoked. There were only two major social divisions: those who were accepted and those who were tolerated. No one was completely ostracized, no matter how socially inept or homely, and there were no bullies. As a result of the fact that both communication and travel were much less developed in those days, being from the big city of L.A. provided me with a kind of mystique, which I used to every possible advantage. I even convinced some of the kids to listen to Harlem Matinee, a very hip black rhythm and blues station in L.A. I had a few infatuations and sexual encounters during the time that I attended this school, but none of them resulted in significant involvement, primarily because I was stuck at the gas station during evenings and weekends and had insufficient extra money for a car.

Maurice Kennedy, a boy who lived nearby, soon became my best friend. He also attended the high school in the next grade above me. Sparks flew when he met Elsie, who had blossomed into a pretty teenager, and it was not long before they were going steady. She became pregnant midway in her second year at Randsburg High, which led to their being married much sooner than they planned. (The marriage stood the test of time; I recently attended their large 50th wedding anniversary celebration.) During her pregnancy, Elsie stayed with a friend in L A. She returned with their baby, Leroy (later shortened to Lee), shortly after Maurice's graduation. Maurice soon got a job at the mining company in Atolia and they set up housekeeping. They had another son, Danny, while they lived there; their third son, Todd, was born later, after they moved to LA. Elsie eventually finished high school and attended college earning an Associate of Arts (AA) degree with honors from Santa Monica City College. At the time of writing, Maurice has been in retirement for more than a decade and they live in Independence, California. They travel constantly, both in the US and throughout the rest of the world. Lee now lives with his wife, Lynn, and their daughter, Allison, in Simi Valley, California. Danny lives in Van Nuys, California. Todd lives with his wife, Sandee, in Westchester, California.

Some months after we moved to Red Mountain, authorities from San Bernardino ran the prostitutes out of town. A few however, returned, and they stayed for several more years. The town remained pretty wild but was much quieter. The service station where I worked on weekends and at night on schooldays was across the street from one of the two town bars, where there were brawls nearly every night. In addition to the action at the bars, all sorts of psychos drifted through town during this period. People would often shoot from the highway, and one night a bullet hit our house, lodging in a windowsill next to where Maurice and Elsie were sitting. After observing evidence of previous activity outside some of the windows, I often slept with my semi-automatic .22 caliber rifle, and it was loaded.

By the second year at Randsburg High, I had regained sufficient strength to play on the school football team; I played for two years. Our team was in the Desert-Inyo league, which included such teams as Barstow and the high school at Edwards Air Force Base. Face guards were just beginning to be used, but we did not have any. I often got bloody from facial abrasions or an occasional bloody nose, likely from the low tackling that I found to be effective. Unfortunately, I never placed much emphasis on athletics, which my dad viewed as irrelevant, but I did enjoy the comradery of participation and the release of aggression.

By this time, I had begun to develop considerable affection for the desert. This was high desert, and on hot days a breeze would always appear in the afternoon and it was pleasant in the shade. Nights in the summer on the high desert are the stuff of poetry; the bright carpet of stars (especially in those days of relatively little light pollution) stirs a kind of primordial, spiritual wonder, and when a light wind stirs the pristine desert air, the most jaded resident pauses to take it all in. I would often go for a walk on moonlit nights after I got off work, even in winter.

During summers after I had regained my strength, I worked days in the mines and nights in the service station. My first work assignment in the mines was the crusher. It was easy to find this machine, which was used to break ore-bearing rock into small stones for further processing; a huge dust cloud hung over it and it emitted a loud, rhythmic, thumping noise, like some enormous, angry monster. As large volumes of rock containing tungsten-bearing ore were blasted from the walls of an open-pit mine some distance away, this material was loaded into dump trucks, taken to the crusher site, and dumped nearby. Bulldozers would then push this material over the top of the crusher. Parallel railroad rails spaced about a foot and a half apart had been fastened to the crusher-top, which was about fifteen feet square. The purpose of these rails was to prevent large boulders from falling into the crusher. As powerful as it was, boulders any larger than those admitted by these rails might stall this monster.

Immediately after a pile of rocks and boulders was pushed to the top of the crusher, I made my way to any boulders that were too large to fall into the crusher, and balancing on the rails to avoid falling in, employed a heavy sledge hammer to break them. Two or three times a day, some of the boulders would be so large that breaking them in this manner would take too long. They then had to be broken up with dynamite. This procedure was initiated by putting blasting caps into each of the sticks before fastening them to the side of each boulder. Insertion of the blasting caps was accomplished by poking a hole into each stick of dynamite, then pushing the cap into this hole. Mud was used to fasten the dynamite to the boulders. The dynamite sticks were interconnected with a continuous wire, the ends of which led to a twist-type electrical detonator. To avoid being hit by flying debris, the detonator was taken a short distance away to behind the shack that contained the main crusher power switch. Although this process was a little scary, it provided a welcome break from the labor of breaking rock. My subsequent employment at the mines fortunately never again involved the crusher.

I often rode around with some of the older guys after I got off work at night, usually drinking and occasionally getting into mischief. On one of these nights, someone had come into possession of a case of dynamite and some fuse and blasting caps. We decided it was time to wake up the town of Randsburg. To maximize the effect, it was placed on the side of a hill facing the town. After lighting a ten-minute fuse, we drove to a friend's house about a mile distant but within the line-of-sight of the dynamite location. A case of dynamite makes a tremendous boom. Even at this distance, it was respectable. I think every light in town came on.

Notwithstanding my frequent late hours, which resulted in napping in some of my classes, I did well in school. One day early in the school year, we were asked to write an essay about some of the material we had recently studied. This class was taught by one of the more competent teachers, a woman who taught history and English. After reading the essay, she sought me out and urged me to go to college. Another event that bolstered my confidence at this time came about when the state issued a call for those who might be interested in a summer job helping on a survey crew. A test for this position was administered in the county seat at San Bernardino. The test included some problems in algebra, a subject for which my training was completely inadequate. Careful reasoning can, however, be used to solve many simple algebraic problems; using my untutored mathematical reasoning, I passed this test. Unfortunately, I was not able to take advantage of this certification, probably because the work was too far from Red Mountain.

Friends often came by the gas station at night. On one of these occasions they told me that there was a big party going on in Ridgecrest, a much larger town about 25 miles away. We went there as soon as I got off work. I was to regret the fact that I did not clean up before leaving. It was a pretty wild and noisy party, and it was not long before the shore patrol – the Navy police - cruised by. (The Naval Ordnance Test Station at China Lake [dry] was nearby.) Some of the kids sensed an impending problem and left. The Ridgecrest police (civilian) arrived soon afterward, taking the rest of us into custody. We were held at the jail in the nearby city of Inyokern for three days. This group of hardened criminals consisted of eleven boys, who were held in one cell, and four girls, who were placed in another. Our small cell held four bunk beds, which we pushed together so we could all sleep. A single toilet without a seat was in the corner of the cell. There were no facilities for preparing food, so we were taken en masse to a local restaurant. The police officer in charge of us actually had his gun drawn as he marched this ragtag group; this must have been quite a spectacle for the other patrons. After appearing before the local judge, we were released to our parents.

My father came to Red Mountain several times while I lived there. He had nearly given up on life before my mom was forced to begin divorce proceedings, and this pretty much finished the job. He degenerated into a typical downtrodden drunk.

David and I had always been close, but when I entered adolescence our social life became totally different. Also, during the time that my mother and I were in Red Mountain, we were usually working. As a result of these changes, David had little guidance or supervision. Consistent with his personality, he managed to get into some trouble, but nothing serious.

My mother moved to Sacramento midway in my senior year, taking David with her. She had always hated the desert. The people who ran the service station across the street had an extra room at the back; I stayed there in trade for part time work at their station. The following summer I worked in the mines, as well as at both service stations. While working at one of the mines, I also assisted in the survey for a new pipeline. Debating whether to join the Air Force or go to college, I chose the latter. At the end of the summer of 1954, I took a bus to Sacramento.

College Years: 1954 - 1959

A combination of circumstances dictated the need for an extra year in obtaining my bachelor's degree. I had to work to support myself and pay for all fees and books, attendance at summer sessions was prevented by full-time summer work at distant locations, I took many more courses than required as a result of a change of major from civil engineering to physics, and I had to take numerous additional courses as a result of the poor training I had received in high school. I never regretted taking the extra engineering courses; most were interesting, and some proved useful later.

The Korean conflict had ended by the time I reached draft age, and after that my grades permitted deferment, so my collegiate career was not interrupted by (peacetime) military service. Years later, I was past draft age when the draft was instituted for the Vietnam War. I was also deferred by my work at that time. This concern about time lost to military service is not without some reservation; I suspect that some of the experiences I missed would have been both interesting and valuable.

The results of the extensive testing performed on entry led me to concentration in physical science and mathematics, the areas of my greatest strength. As luck would have it, the strongest part of the curriculum at Sacramento City College, the first college that I attended, was in the areas of mathematics, physics, and engineering. The physics courses were so difficult that some students avoided them by taking these courses at U C Berkeley during the summer. (In contrast to the high quality of these courses, some of the liberal arts courses were pretty easy, but this was of little concern to me at the time.) Based on entry tests and my performance on an unexpected essay, I was admitted directly to college English, bypassing the remedial coursework in this area that I had expected.

My socialization was somewhat compromised. I felt awkward and out of place. I was a product of another culture, and the responsibilities and experiences of my life to this point left me feeling older than the other students, who mostly appeared to me to be soft and immature. I felt some kinship with veterans returning from the Korean conflict and attending on the G I bill, but most of these guys were busy raising families. I also had to work, which reduced available free time. Nevertheless, I should have attempted to participate in more social activities with other students. My eventual professional life would have benefited from the ability to interact on a more personal level with people whose backgrounds were so different than mine, and I needed to shed some of the negative baggage of my previous environment.

I spent a few weeks at the Sacramento home of my Aunt Blanche and Uncle Blackie before moving to a boarding house located near the center of the city. This place, a short walk from the skid row area, had been converted from a very large old building. Each hallway was rigged with a bell on a string that was pulled at mealtimes. When the bell rang, there was a mad rush downstairs to the long dinner table. The noise and vibration produced by so many running down the old stairs seemed to threaten the structural integrity of the aging building. An abundance of cheap starchy food was served, and self-preservation led to the rapid development of a good "boarding house reach". Although the skid row area could be dangerous at night, I used to enjoy walking down there early in the morning, where I could get a doughnut and a cup of coffee for a nickel. The tenants at the boarding house ran the gamut from semi-permanent residents to some pretty low-life and transient types. My only suit was stolen there. I moved out after a few months.

My funds were running low and I got a job at a service station, temporarily boarding at the home of the family who leased it. I got my first car at about this time: a 1948 Oldsmobile; this served me well for several years. I often sat in my car outside their home, listening to the radio, procrastinating my studies, and enjoying the beauty of Sacramento in the fall. The older part of Sacramento is reminiscent of many eastern towns. Much of the larger buildings are constructed of brick, and the bells of many large old churches toll the hours. The city is filled with deciduous trees, providing an abundance of color in the fall.

The boarding arrangement soon became strained, so I quit the service station job after I found a part-time job at a veterinary hospital that included a room. The room, which was at the back of the hospital, was about seven by eight feet and held a single bed, a lamp, and a small gas-operated heater. The heater was barely adequate and the room was poorly insulated, so as I sat close to the heater while studying at night, my back would get so cold that I had to pause frequently to turn around and warm my backside. In an amusing incident there, I was awakened one morning by a large dog atop my bed, standing over me and panting in my face; he had escaped from one of the cages during the night.

The animal hospital was located near the apartment where my mom lived; on weekends, I would go there for lunch. She was working at a ceramics shop, which she enjoyed, and she had lost a great deal of weight. She was happier than I could remember, and she looked wonderful. She had apparently reached a major turning point in her life. Although she later gained the weight back, most of the old sadness and anger was gone, and thereafter she remained more content and at peace with the world.

I was homesick and I had never needed to study prior to entering college, so adjusting was difficult and I considered the possibility of dropping out. At the end of the first academic year, I returned to the desert to work in the mines. A summer of back-breaking labor soon renewed my resolve to finish college.

Returning for my second year, I rented a room with kitchen privileges in a very nice part of town; I was to live here for much of the next three years. The older lady who lived there and rented out two of her rooms became a good friend. The house had a small sitting room, and when I was not studying or working, I would often watch TV with her there. (Nearly everyone had a television by then, and although the pictures were black and white, the programs had become much better and the screens were bigger.) During much of this year I had a job changing marquees at movie theaters. The only bad parts of the job were being called out on stormy nights and changing the marquee for the theater on skid row. Quite often when I was changing the skid row theater marquee late at night, drunks would stumble into the ladder, with the possibilities that I might lose my balance when stretching to reach a letter, or that the ladder might be knocked over. The least injury that I could have expected from a fall of that height would have been a broken leg.

Although I had done well in my first year, there was room for improvement. I really buckled down in the important classes during the second year. This is one of the few large-scale undertakings in life where the influence of politics is minimal, and reward is very nearly proportional to effort expended. Thanks to diligent study, I led nearly all my math and physics classes. The fact that I could not quite surpass one particularly gifted individual in two of these classes provided a valuable lesson in humility.

Between my second and third years, I got a summer job on a mountain survey crew that was assigned to establish topographic mapping for a proposed road through the coastal redwood forest. In open country, these maps are commonly accomplished by use of a type of stereo photography mounted on aircraft, but when trees obscure the ground, this method cannot provide the accuracy required. The area we surveyed was near an approximately twenty-mile stretch of old Highway 101 in the vicinity of the towns of Rio Dell and Meyer's Flat. This is the part of the redwoods containing the world's tallest trees. In some places, direct sun is not seen until midday. The beauty of these areas has a grand but somber quality, evoking feelings probably best described by the music of Beethoven. In order to perform a topographic survey, the surface of the ground must be in view. A kind of optical sight is mounted atop a table on which the map is drawn; the term "plane table survey" is sometimes used to describe this work. Sightings are taken on special rods held at carefully chosen locations by members of the team, allowing the determination of both distance and elevation relative to the sight, from which contours of elevation are determined. The terrain was rugged and the foliage dense, but the reward of working through it was to experience some of the most pristine scenery imaginable. Sometimes after making our way through dense woods, we would find ourselves in a clearing with huge ferns of a lighter, brighter green, contrasting with the dark forest. We once came upon an area where a soil slippage had taken place, severing the roots of a large stand of trees below ground level; the trees were so close together that they held one another up, even though most were dead. The scene inspired a feeling akin to reverence.

At first, we were based at the town of Rio Dell, where I became involved with a local girl. Rock and roll had finally brought some life to the drab music repertoire of white teenagers, and we spent many hours necking to the music of new performers like Elvis Presley. Toward the end of the summer, the survey team was moved to Meyer's Flat, where we stayed in a bunkhouse next to an old hotel. I liked this work and was urged to stay on; it was tempting, but I declined.

When I returned to college in Sacramento, I was able to rent the same room in the nice part of town. I soon got an evening and weekend job at a trailer rental agency. By this time, my mom had met and married Lorin Rogers. They had a restaurant in Loomis, some distance north of Sacramento, where I occasionally visited on weekends. Lorin was a cheerful and stable non drinker. My mother became a different person than when she was married to my dad; anger and pessimism had been replaced with optimism and a sense of well-being. My half brothers, Ralph (later changed to Ray) and Mark were born at about this time;

they both inherited their father's sunny disposition. At the time of writing, Ray has retired from the Navy and has just completed his MBA; he lives in San Diego with his wife, Akemi, and son, Greg. Akemi's son by a previous marriage, Bret, moved out after obtaining his Bachelor's degree in physics; he is employed in work related to his training and plans to attend graduate school. Mark has the enviable position of liquor manager in a supermarket; he lives in Orangeville, California with his wife, Shalois, and sons, Scott and Keith.

I stayed in Sacramento the next summer, where I continued to work at the trailer rental agency. Sacramento gets very hot in the summer. I would often get together with my brother, David, who was in high school, and we occasionally took girls to the American River near the town of Auburn. This is a mountainous area covered with a dense coniferous forest. Many extremely large boulders are to be found on the riverbank; we would bask in the sun, lying on one of these boulders above deep, dark cold pools in the river. David later joined the Coast Guard, eventually settling in Oregon. He is presently living in Colorado with his son from his second marriage, Ray, near his youngest daughter from his first marriage, Debbie, and her family. Diane, his older daughter from his first marriage, lives in Oregon. Lacey, his daughter from his second marriage, is in the Air Force, stationed near Sacramento.

By this time, I had settled on a major in physics with a very strong minor in mathematics. Having finished my freshman and sophomore years, I had to decide where I would go to complete the requirements for my bachelor's degree. Engineering and science student transfers from Sacramento City College were well respected in the upper divisions of Stanford and Berkeley, but my relative penury meant that Stanford was out, and physics majors at Berkeley had to satisfy strict language requirements in German and/or French. My limited experience with these languages had convinced me that progressing beyond the conversational stage represented a nearly insurmountable obstacle. I struggled at length with the issue before deciding to complete my college years at the California State University at Sacramento, reasoning that the degree meant employment, and I might later study for an advanced degree either in some branch of engineering, or at an institution with a less demanding language requirement. As luck would have it, Cal State U Sacramento had an excellent mathematics department. The physics department, however, was less developed. The professors were competent and wanted to expand the department, so it was easy to influence the material covered to ensure that I was gaining as much background as I would expect to receive at any institution. I really delved into it, working nearly every problem in each of the various textbooks; as a result, I led all these classes by a substantial margin. I have many pleasant memories of this time. In the spring I would often sit with friends on the huge central lawn of the campus, where we would attack the kind of philosophical questions that have always been discussed by young people. This also provided the perfect opportunity to watch the coeds walking by in their spring dresses.

During the last summer break, I worked at the atomic test site at Mercury, Nevada. I had a comparatively menial job there as a clerk and dispatcher for the mobile communication network. I also learned to run the big carbon-arc projectors at the base movie theater, where I often filled-in at night. I obtained my first security clearance there; everyone who worked at this site had to have a "Q" clearance. I never suspected that I would return many years later in a far more important capacity.

When I returned to Sacramento, I was unable to rent the same room. I found the cheapest rental available, which was in the basement of a boarding house located downtown. It was dismal; I had to walk in a crouched position over boards placed on the dirt floor under the building to get to my room, which was a small hovel lined with concrete. Fortunately, there was one window, about six inches high, at ground level. About halfway through the year I got a couple of jobs, and David and I moved into a small rental house a short distance north of Sacramento; we had some pretty wild parties there.

As a result of their familiarity with my capability, the faculty of the physics department took the unusual step of hiring me in a teaching position, even though I was still an undergraduate. My job title was lecturer in physics; I taught the laboratory section of college sophomore electricity and magnetism to engineering and science majors. I took the job very seriously, rewriting many of the experiments. My other job consisted of part time work at a service station. Between teaching, working at the gas station, and carrying a heavy load of mathematics and physics courses, I was so tired that I slept through my graduation (I slept through two alarm clocks; one some distance from the bed). David bought me the very nice graduation gift of a leather-bound portable radio that was large enough to produce great sound, but small enough to be easily portable.

The degree in physics, supported by nearly enough credits in mathematics for a double major, was conferred in June of 1959. I had been elected to Pi Mu Epsilon, the national mathematics honor society, the previous year. Sigma Pi Sigma, the national physics honor society chapter, had not yet been installed at this institution, but I was also later elected to this society. Aerojet General and Douglas Aircraft Missiles and Space Systems

Division recruited me, and I received unsolicited offers of assistantships for graduate study from USC, UC Santa Barbara, UC Davis, and the University of Arizona. I was sick of being poor and I reasoned that I could always study for an advanced degree later; also, the type of problems most physicists were involved in at that time seemed less interesting than missiles and rockets. I decided to accept the offer of Douglas Aircraft.

I pursued additional studies throughout my career. Beginning shortly after I joined Douglas Aircraft and was transferred to Santa Monica, I took courses in electrical engineering at UCLA including communication systems, digital circuits and logical design of computers. These courses, which were more directly related to the work early in my career than was physics, were later interrupted after my place of employment changed, requiring a long commute. Subsequent company-funded studies included accelerated short duration courses. Two of the best of these were data compression at MIT in Cambridge Massachusetts, and fiber optic transmission systems at the Naval Postgraduate School in Monterey. Because most of my career was spent in research and advanced design, which often surpassed existing course material, self-study of both texts and the latest scientific papers was frequently required.

First Marriage, Children, and Early Career: 1959 - 1970

I hired into the Missiles and Space Division of Douglas Aircraft on June 24, 1959. I suspect that some of those holding bachelor's degrees from top schools like MIT or Cal Tech were likely offered a few percent more than those with bachelor's degrees from other institutions, but within a short time, one's actual knowledge and demonstrated capability became the only relevant performance prediction metrics. After that, the question of alma mater never surfaced. I quickly overcame any initial institutional bias, as well as the starting premiums paid for advanced degrees. By career midpoint, my salary exceeded that of many of my colleagues with doctorates. (Denigration of the attainment of advanced degrees is not intended. The possession of uniformly recognized certifications of academic achievements is always a plus; also, I enjoy learning nearly everything, and working toward these milestones would likely have included studies in areas that I have neglected.) I used most of the extra time that could have been spent fulfilling the requirements for an advanced degree on directly relevant self-study and additional uncompensated evening and weekend work. Although I took numerous company-funded courses related to my work in the beginning of my career, as well as a few much later, much of my work rapidly changed to areas so new that no university courses were available. As a result, preparation for involvement in a different area often could only be accomplished through study of the most recent work contained in scientific journals.

In my opinion, the primary advantage of working for a large, diverse engineering company is variety. In those instances where a particular program ended, or I became bored or misplaced, other fundamentally different work was always available. In effect, I worked for many different companies. By the time my career ended, I had worked in the areas of data transmission, signal processing, electromagnetic nuclear effects, sonar, electro-optic sighting systems, and certain classified systems. This work, which primarily consisted of research, initial design, and analysis, entailed the acquisition of substantial concomitant knowledge in space, atmospheric, and ocean sciences, as well as in certain branches of mathematics and engineering. I estimate that I spent at least a quarter of my career in the study of scientific papers and texts. Although individuals associated with implementation, deployment, operation and maintenance tended to travel more than those, like myself, who were involved in research and design, I traveled considerably more than most. Fortunately, the majority of my destinations and off-site activities were both challenging and interesting.

My first assignment was in the area of missile test data processing at the Sacramento field station, where test firings of the Thor missile were conducted. The rocket engine used for these missiles was liquid fueled. The oxidant was liquid oxygen or "lox", as it was commonly called; this is dangerous stuff: if it touches the skin, the tissue effectively catches on fire. Although I was about a hundred feet away during this process, the adrenaline level was still elevated. The sight of the missile being fueled was impressive. Lox is extremely cold, so a cloud of vapor surrounded the missile and large chunks of ice fell from the sides. The firing itself could only be viewed at a much greater distance.

David and I moved to a small guesthouse at the rear of a property in Roseville. With my newfound and disproportionately inflated sense of wealth, I soon got a loan for a black 1957 Corvette convertible with a red interior. I loved driving with the top down; when I drove to work in the morning, I did not put the top up until the temperature dropped to below freezing. Although I had been driving for many years, the first time I took a couple of girls for a thrill ride I almost rolled it. One night while driving with David I pegged the speedometer at 140 mph; it had no roll bar and no seat belts.

I was transferred to the Santa Monica facility of Douglas Aircraft early in 1960. During the eight-month period between arrival in Santa Monica and the time of my marriage, I boarded at the Santa Monica home shared by my Aunt Gertrude and Grandmother Emma. (My Uncle Raymond had recently moved out.) My grandmother was a redhead with a nasty temper; we had a couple of spats, but otherwise got along very well. Her marriage with my grandfather before they permanently separated must have been interesting; like her, he had a bad temper. I heard that he once lamented that she no longer cared for him because she would not fight with him. It is not surprising that tempers run high in many of their descendants. At the time that I lived there, Emma had retired from teaching, and had been painting for some time. I used to ferry her, together with selections of her paintings, to various open displays where artists sold their work. She loved riding in the Corvette with the top down.

The processing of large amounts of data required extensive electronic systems. In support of my work in this area, I began taking company funded electronics engineering courses at UCLA. I was so enthusiastic and hard working that some of the managers began talking to me about moving into management; I tactfully downplayed the issue. I had seen the disastrous results of young managers who lacked adequate experience and knowledge.

As stated in the Genealogical Information section, I married Doris Mauro in September of 1960. I had met her shortly after transferring from Sacramento eight months earlier; she was a data clerk in the same area where I worked. Doris was an attractive 22 year-old (I was 24) with the unusual and striking combination of red hair, brown eyes and a beautifully tanned complexion. Her background was fascinating: she was born in Italy and did not arrive in America until the end of World War II. Her slight accent was also interestingly unique. At the time, I had a girlfriend from Sacramento who had also transferred to Santa Monica, so I had to endure the miserable process of ending that relationship before the courtship with Doris could develop. Of course, she also had to end her involvement with another boyfriend. Our honeymoon began in Mazatlan, Mexico, before it became a popular tourist spot; from there we toured Mexico City, Guadalahara, Taxco and Acapulco.

Shortly after our honeymoon, we moved into an apartment in Santa Monica near Doris's sister, Evelyn. This was not a good idea; Evelyn was too dependent on Doris and constituted much too great a presence in our early marriage. (Evelyn's [and Doris's] father, Joe, and Evelyn's son, Danny, also lived with Evelyn.) Our first two children, Lisa and David, were born when we lived there. I attended my first out-of-state conference, which was held in Boston, during this time. Lisa could already pronounce words well, and we got a kick out of her flawless pronunciation of "Massachusetts".

We moved to our first home, which is located in Mar Vista, about a mile from Evelyn's apartment, shortly before Monique was born. My brother's daughters by his first marriage, Diane and Debbie, and my sister's last child, Todd, were also born in this time period. The kids were the center of our universe; some of my happiest memories are of their childhood and the experiences I shared with them. We did not spoil them, and while discipline was never withheld, it was never administered without complete communication.

In early work I was able to characterize a particular large-scale electronic system with a set of equations that could be used to determine areas in which frequent component replacement would substantially reduce down-time. I documented this work, like much of that throughout my career, as a stand-alone monograph. (Many of these documents were distributed to numerous industry and government factions, and a few formed the bases of my formal publications.) As a result of the recognition I received for this work, I was placed in a research group. The primary focus of my activity soon became the emerging discipline of digital signal processing, for which no texts were yet available. This was true of much of the work throughout my career; I often worked in emerging technologies for which only a few papers were available. At that time, I also worked on various systems for telemetry, the science of the efficient transmission of large amounts of data from missiles and spacecraft; this work entailed attendance at a conference in Houston, Texas. My primary memories of this city are of great steaks but merciless humidity.

Luckily, my work at that time afforded me the opportunity to publish, and I presented my first major international conference paper in Washington, D.C., in May of 1965. This paper, "A New Approach to Effective Digital Filter Design", was one of the earliest in this field. (If interested in an abstract, Google: P H Duncan effective digital filter design). My later papers underwent much more rigorous peer review and had far wider distribution. I was twenty-nine years old. Like most of my later publications, this represented an enormous amount of work conducted over an extended period, the final product of which contained nearly as many equations as words. Assistance in the process of preparing to make this presentation was provided by both a company speech coach and a boyfriend of Aunt Gertrude's, who taught public speaking at a local community college. It was a long speech - about forty-five minutes. It was also the first of the conference, so the attendance was high - about two hundred people. I learned that the secret to a good presentation was rehearsing, preferably five complete times without error. I had typed the entire speech for backup, but never needed to refer to it once during the long speech. The fact that the audience's attention was usually directed to my charts helped a great deal. I was required to give presentations from time to time throughout my career, and I usually followed the same procedures of conducting at least one complete and several partial rehearsals and directing much of the attention to charts. Although these carefully prepared presentations were virtually always well received, I regret that I never became an accomplished extemporaneous speaker.

Inevitably however, I occasionally had to speak without preparation, and the results were acceptable, but I was never comfortable with the process.

Later in 1965, I was relocated to a new facility in Huntington Beach, California, where I worked on the Manned Orbiting Lab program. I already had a secret clearance, but this was soon augmented by crypto and compartmental security clearances. During this period, Douglas Aircraft merged with McDonnell Aircraft to form McDonnell Douglas. My work included trips to Edwards Air Force Base near Boron, California, the Martin Aircraft facility in Loveland, Colorado, and the new company headquarters in St, Louis, Missouri. The trip to Edwards was unusually interesting because we had to hold our flight pattern (we were in a light plane) for about 20 minutes to allow the landing of an X15 manned rocket. We were too far away to see much of the rocket, but the dust plume, which appeared to be more than a mile long, was impressive and we were able to monitor communications with the rocket. The trip to Loveland provided a convenient way to see my dad, who lived nearby in Denver. During this visit, we drove to Nederland, where I saw the tiny house where I was born for the first time. He was still drinking but looked much better than during the desert years.

I soon tired of the commute, and in 1967 I joined a Santa Monica based research group within the company. During this time, I took a company funded course in the emerging field of data compression at M.I.T in Cambridge, Mass. In 1968 I completed a short book summarizing my work and other research findings in the area of data compression. The management thought some of this work was particularly useful, and I was taken on an extensive tour to present some of these findings to various military factions. This tour included Washington, D.C.,

Fort Monmouth, New Jersey, and Rome Air Development Center in Rome, New York. The flight to Rome, in a Fairchild turboprop, was interrupted when our altitude was quickly dropped by about fifteen thousand feet and the cabin was depressurized; my ears ached for a couple of days afterward. This event forced an unplanned landing at a snow-covered airport in White Plains, New York. It turned out that the cockpit window had begun to shatter, necessitating the depressurization. We rented a car and drove the remainder of the way to Rome.

When the kids were about three years old (Lisa turned three in 1965), we enrolled them in a pre-school co-op nursery school; this proved to be an extremely positive experience for all of us. The families taking part in this co-op were unusually diverse, but they all shared the desire to raise children that would become well-adjusted and productive adults. The teacher was one of the most interesting people I have ever known; she was a well-educated but creative and objective thinker with an admirable ability to work with the ideas of others. Many group meetings with this teacher and the parents were convened to discuss parenting issues. The primary theory advocated was that of natural consequences; for example, if a child hits you, the natural consequence is to hit him back, because that is what might well happen in the adult world. In the numerous cases where a natural consequence is not practical, a generally moderate but firm approach was advocated. I had seen the undesirable consequences of the extremes of permissive or authoritarian parents, so I found most of the ideas fairly logical and practical. The kids blossomed there, forming many lasting friendships. I thought my children were the cutest, of course, but they had a lot of competition in this category from the other kids in the school.

Holidays, birthdays, and graduations were always occasions for lavish celebrations. Doris and her sister, Evelyn would prepare an elaborate dinner in accordance with Italian tradition for midnight of Christmas Eve, an event that always drew a crowd. Our good friend Ann Stevens prepared the dinner on Christmas Day; this dinner was also well attended. I would stay up late the night before Easter, coloring eggs and hiding them in the back yard; the tradition of the hunt for them on Easter morning continued well into the kids' teen years. Birthdays also occasioned a big bash; after a big afternoon party, the adults would continue the party well into the night. In later years I would set up a bar outside, making frozen margaritas by the gallon.

The late sixties and early seventies constituted the era of the Vietnam War and associated protests, civil rights marches and protests, and free love. The uniform of the day consisted of wide collar shirts, wide belts, and tight bell-bottom trousers. AIDS did not arrive on the scene until more than fifteen years later, birth control pills had become available, and women went braless and wore miniskirts. It would have been a great time to be single, provided you were also unavailable for the draft. There was a dark side, however; heroin addiction was common and much of the music of the era had a dark quality, reflecting the undercurrent of a world that seemed to have lost its bearings.

Before long, the research group was relocated to Huntington Beach. One of my assignments was to perform a brief study to determine the expected future direction of data storage technology. A method under investigation by a company in Redwood City, California, seemed to show promise, and I traveled there to meet with their research people. Unfortunately, however, much of the work of the group at that time was in areas in which I had little experience or knowledge. The negative effect on my career was compounded by massive cuts taking place in the space exploration industry after the moon landing, and my own depression and some drinking issues. Also, Doris and I had taken on the responsibility of trying to bring some direction into our nephew (Doris sister Evelyn's son) Danny's life. He was a teenager by then, and he was often getting into trouble; after he moved in with us, the situation eventually degenerated to the point that he and I once got into something of a scuffle. I regret that my problems at that time severely limited my ability to help him, but he eventually recovered and did very well. Danny now lives in Los Angeles with his wife, Lynette, and their two children, Kyle and Skylar; they have a large home there, which they have extensively refurbished.

If the downward trend in my career had continued, I would likely have been out of a job. (Seniority is virtually irrelevant in the engineering departments of most, if not all, companies of this type. For example, given the option of keeping only one of two engineers of otherwise equal value to the company, the one with the least seniority will usually be retained.) Those with political ability who use much of their time to ensconce themselves in a "good old boys" club containing some highly rated members might delay termination, but for nonpolitical people like me, duration of employment per se is of little or negative relevance. I began interviewing various factions at the company to find a new position. The most interesting interview was with a crusty, older engineer who was assembling a team to perform a four-month nuclear-effects test on the Spartan anti-ballistic missile (ABM) missile at a facility near Albuquerque, New Mexico. This area of work seemed interesting, and I welcomed the time away. He agreed to take me on to lead part of this team. This was the beginning of one of the most exciting periods of my career.

Mid Career Period: 1970 - 1990

The first nuclear-effects test performed on the Spartan Missile was conducted during the winter of 1970 - 71, in a test facility that was built to simulate nuclear generated electromagnetic pulses. The interaction of blast-generated electrons with the earth's magnetic field produces an extremely powerful electromagnetic pulse (EMP). The EMP resulting from a high-altitude burst can travel halfway around the world, causing massive disruption of electrical and electronic systems, including those contained in an anti=ballistic missile like Spartan. The test facility replicating this pulse is located in Kirtland Air Force Base, near Albuquerque, New Mexico. A multi-million-volt generator produces the pulse, which is transmitted to the test volume by a huge wave guide system composed of a network of cables. The response of missile interior systems to the pulse was routed via shielded cables to a bank of camera-mounted oscilloscopes. As data group leader, my primary responsibility was to define the best settings for the instrumentation before each "shot", and identify, interpret, and catalogue the information obtained.

The test team consisted of about half engineers, half technicians, and representatives from the warhead manufacturer and Bell Telephone Company, the prime contractor for the complete ABM system, which also included the faster, shorter range Sprint missile. (The basic concept of this missile defense system was to take out incoming ballistic missiles with maneuverable three-stage Spartan missiles, then destroy any that got by the Spartan missiles by use of Sprint missiles.) The test was also periodically monitored by the Air Force. We had Sunday and an occasional Saturday off. We worked long days and drank a lot at night. There was a great deal of esprit de corps among us, and whenever we got together with the old boss, a heavy scotch drinker, the talk was always of the work.

On one of the weekends in Albuquerque, some of the technicians and I decided to rent some dirt bikes. Some of these guys had these motorcycles and one of them was a motocross race driver. I made the mistake of trying to keep up with him; near the end of the day I had an accident resulting in an extremely painful type of shoulder injury. The usual procedure for treating this injury was to operate and put a pin through the shoulder socket. Since this generally resulted in problems later-on, a new semi-experimental procedure was used in the hope of reducing long term problems. I was put in a kind of traction device that had to be worn for six weeks. I lived on pain pills but never missed work. As soon as I was released from this contraption, I began lifting weights. The shoulder bothered me for several years afterward but was eventually as good as new. I continued lifting weights for many years, later adding jogging to my routine. At age forty-seven, more than a decade later, I could still bench press 300 lbs even though I had never trained to lift heavy. (My body type is somewhat athletic in appearance, similar to the Pearson men, who are thickly built and 5 feet 8 or 9 inches height.) I normally weigh about 160-165 before I quit smoking to 170-175 lbs. after, and my barefoot height at maturity was about 5 feet 9. As a result of spinal compression, I lost height rapidly beginning in my fifth decade, eventually dropping more than two inches.

Albuquerque is located atop a large mesa. The skies are a deep blue and a mountain resort is nearby at Sandia Peak. The forested area in the foothills contains a mix of deciduous and evergreen trees and the drive to the top is striking in the fall. The company put each team member in a new condominium, and we all had rental cars throughout our time there. Doris and the kids visited me there a couple of times; they thoroughly enjoyed the area.

On one extended holiday weekend, I drove to Denver to visit my dad. He had visited me at my home in Mar Vista during a non-drinking period a few years before, but by this time he had started to drink again, although much less than before. He had a part-time job and must have been making a little extra, because he was feeding every cat in the neighborhood. This turned out to be the weekend of one of the worst blizzards ever, but the sight of the snow contrasted with exposed cliff sides of red stone in the rugged country of northern New Mexico made the decision to drive worthwhile.

The missile was tested in various orientations. When, while tested vertically, it was first turned on its axis by 180 degrees, responses increased dramatically, and it was feared that this might be the result of extraneous pick-up by the instrumentation cables, rendering the test invalid. The engineers on the team, as well as consultants from the firm who built the facility, were cloistered in a large on-site trailer with a mandate to determine if the test could proceed. After nearly two weeks of head banging, I finally came up with the explanation that validated the readings. Subsequently, a team was sent from headquarters in Huntington Beach to review the test, especially this last development. The team included a well-known expert. When I explained how I solved this problem, he was so pleased with it that one would think that he had solved it. He had a pronounced but not pompous British accent, and on most of the few occasions when we happened to cross paths over the years, he greeted me as his "distinguished colleague". While this would seem odd and humorous to our American ears, I knew it was merely a friendly greeting by a very British colleague acknowledging our membership in the same "club".

At the completion of this important and successful test, the team received written commendations from all the factions involved. I wrote a significant part of the final report, which was widely distributed throughout relevant government and aerospace factions.

I returned to Kirtland AFB several times over the next two years. The first was to perform a gamma ray test on the Spartan third stage. This was performed at a different test facility than that used for EMP testing. In addition to EMP, a nuclear burst is also the source of an intense pulse of gamma rays. These rays interact with the missiles electronic systems in various ways, one of which is the generation of a large amount of current in the form of a short pulse of free electrons within the missile. This is referred to as internal EMP, or IEMP. This is the phenomenon we were measuring with the test in this facility, which employed an immense device to produce the required pulse of gamma radiation. In one interesting incident, we were measuring certain properties of the radiation field using an array of materials and measurement devices designed to accomplish this task. I had a serious disagreement with the chief nuclear physicist from headquarters in Huntington Beach regarding the placement of these materials. Probably much to the amusement of other team members, the debate raged long into the night before the test, without resolution. When I arrived at the test site next morning, those who were already there were looking at me with quiet smiles. Puzzled, when I entered the test volume, I found that this physicist had already arrived and placed the materials and measurement devices exactly as I had argued. I gracefully kept my mouth shut.

By this time I had become well established in the nuclear-effects community. I soon traveled to Bell Telephone Labs in Cherry Hill, New Jersey, for consultation on EMP testing. Later, I was asked to participate in the national nuclear effects working group held at the Antler's Plaza hotel in Colorado Springs, Colorado. I particularly recall the flight from Denver to Colorado Springs in a DC 3. This aircraft is the civilian version of the legendary C 47 used to drop airborne troops in WW II. After becoming accustomed to jets, and having experienced flight in a small plane, I was amazed at the amount of vibration the large engines transmitted to this aging aircraft's structure while revving up for the extra-long take-off run necessitated by the airfield's altitude.

As a result of our successes in the Spartan tests, we were asked to perform the gamma ray IEMP test on the Sprint missile, which was built by the Martin Company in Orlando, Florida. The Sprint missile is much smaller than the Spartan, so the test was conducted at a facility with a smaller gamma ray generator than that at Kirtland. This generator was located at the Harry Diamond Laboratories in Washington, D.C. It was late fall, and this beautiful city is filled with deciduous trees, so there was color everywhere. The test required a little over two weeks to complete. The company always put us up at large, first-rate hotels, and our travel was well compensated, so there was a lot of drinking and fine dining.

During this era, Congress authorized the formation of a scientific committee to monitor the wide variety of nuclear effects testing that was being performed by contractors all over the U.S. This group of distinguished scientists was known as the Shea committee, after John Shea, the leader of the committee. I was chosen to present the results of the Sprint gamma ray test to this body, and the presentation was to include extrapolation of results to the response expected in the presence of an actual nuclear burst. This was not easy; since gamma rays generate electrons within the measurement system, the response of the missile's electrical systems was often buried in a great deal of noise. My task was to extricate the missile's response via extensive and carefully reasoned analysis. Since the burst of gamma rays from a nuclear detonation is much greater than that which could be simulated with the generators, the extrapolation to threat levels was extremely difficult. I was understandably nervous about the presentation. This apprehension was significantly increased when I arrived in Washington, D.C., the evening before my presentation. Meeting with others who had been at the presentations given to the committee that day, I learned that the committee had brutally criticized nearly all the presenters. I was to be the first presenter the next morning. I didn't get much sleep that night, but my extensive preparation paid off and the presentation came off without a hitch.

To gain a better understanding of the processes involved in IEMP, it was decided to perform additional testing via a method known as current injection, which consisted of the simulation of internal missile electromagnetic fields by electrical methods. The Spartan part of these tests, which were performed in a facility at Santa Monica, employed a metal model of the missile. The difficulty of the test was compounded by the extraneous presence of a consultant from an electromagnetic consulting firm on the East Coast. A young man who had recently completed his doctorate with emphasis on electromagnetic theory, he was irritatingly arrogant, and our irritation with him was increased by his habit of employing the feigned accent of a "southern gentleman". Although he was uniformly disliked, the conflict between he and I was much greater than that with the rest of the group. After proving him wrong in arguments involving electromagnetic theory, (one of my favorite subjects), the tension between us increased. He could not understand how I, without a doctorate, could understand this much theory. (This always seemed strange to me -- many seem to think that attendance in classes is a mandatory requirement for the study of theoretical texts. Of course, the advanced degree encompasses a much broader and more balanced range of related topics.) Things finally came to a head one night when we all went to a bar to unwind. In his usual irritating manner, he was bragging about his prowess at karate and acting like an arrogant, all-around jerk. My irritation with him was probably enhanced by my nicotine withdrawals; although I had quit smoking many months before, I was still occasionally irritable. At some point my body simply took over and, out of character, I took a swing at him. Fortunately, the combination of his evasion of the punch and my alcohol level resulted in a grazing miss. The situation was rapidly defused, and that was pretty much the end of it, but thereafter he addressed all of us with greater consideration, if not respect.

Soon after completion of the Santa Monica current injection test, I traveled to Orlando, Florida, to assist in current injection testing of the Sprint Missile. Because this test involved access to a warhead sans fissionable material, I had to add a critical nuclear design information (CNWDI) clearance to my collection.

After returning from Orlando, the chief nuclear physicist for our activity advanced the concept of reducing IEMP effects by filling much of the interior of the missile with a lightweight conducting foam. We constructed a test specimen, which I tested in a small gamma ray facility located in Palo Alto, California. Stan (the physicist who proposed the concept) and I collaborated on a paper describing the results, which I subsequently published and presented at a national conference held at Kirtland AFB. This classified paper: "Attenuation of EMP/IEMP Effects Through Low Density Foam with High Electrical Conductivity", was published in the (classified) proceedings of the 1973 National EMP Conference in Albuquerque, New Mexico.

At about this time I attended a nuclear effects conference in La Jolla, California. While sitting in a particularly boring session, I took the opportunity to scope out an analysis I had been thinking of tackling for some time. After publishing the completed work sometime later in an EMP/IEMP journal, it was spotted by the editor of a world class journal in a more general related field, where it was published verbatim, effectively speeding through the usual lengthy peer review process. Although my national conference papers had achieved archival status via publication in the proceedings of major conferences, this world class journal was significantly more prestigious; the overwhelming majority of those whose work is accepted for publication in journals at this level hold doctorates. This achievement was a tremendous status builder among my peers. The paper: "Analysis of the

Mobius Loop Magnetic Field Sensor", was published in the May 1974 issue of the Institute of Electrical and Electronic Engineers or IEEE Transactions on Electromagnetic Compatibility. (The IEEE is the world's largest organization for the advancement of technology.) If interested in an abstract of this paper, Google: Paul H Duncan ieee loop. I later published two additional papers in IEEE journals.

When returning home from a Christmas party held in a colleague's home at about this time, I failed to pass a roadside sobriety test; I was cited for DWI and taken to the main Orange County jail. At the jail, just like all the other individuals they had picked up, I had to strip and stand under a shower in a large concrete room, after which armpits and crotch area were sprayed with disinfectant, and I was issued jail clothes. I was then put into a cell holding about 30 individuals. This cell had three metal toilets without seats, and as many showerheads; they were not separated from the remainder of the cell, so there was no privacy whatsoever. The following morning, guards ushered us into the jail cafeteria where breakfast was served on a metal plate. I was bailed out later that morning. The experience was a fortunate one, for it forced me to deal with my alcohol issue. Because I enjoy a glass of wine or a beer with dinner, and occasional, sensible social drinking, I did not abstain completely. With some practice, I found that I could easily restrict my intake of both regular and occasional use.

Doris and I took an extended tour of Europe during this time period; this tour began at the town of her birth in southern Italy, where we stayed for about three weeks. Part of the time was spent with her mother's relatives. When the supply of wine ran low, Doris' cousin and I loaded two jugs about the size of those used in water coolers into his tiny car, and we headed into the nearby mountains for wine tasting. This procedure was a little different than that employed here. Each vintner had a huge barrel, often representing his entire small grape harvest. Tasting involved half a water glass, so the tour was a short but merry one. The remainder of our time in Southern Italy was spent with her father's relatives, some of whom were quite wealthy. Our next stop was Rome, where we met more of her father's relatives. We were wined and dined at a couple of the finest restaurants in Rome. From there we proceeded via train, van, and plane to Florence and Venice, Germany, Austria, and Paris.

By the beginning of the seventies, the kids were old enough so they could enjoy camping trips. Most of these excursions were either next to the Truckee River south of the town of Truckee, California, or in the Eastern Sierras near Bridgeport. Most of the campgrounds were of the small, pristine type where large motor homes are not allowed. We typically stayed for the duration of my vacation, far from the stress and pollution of the city. The kids loved it, and I loved spending time with them. On one especially long vacation in the mid-seventies, we also took a five thousand-mile tour that included Western Wyoming, a remote area in Montana, and a brief visit to my dad in Colorado. Three active, sweating kids in the back of a sedan sometimes proves exasperating, but we had a wonderful time. Before long I became somewhat fanatical about fishing. At first, Lisa was the most avid fisherman, but David soon got into it. Monique just enjoyed it all. Lisa's interest in fishing eventually waned, but David became progressively more interested. Although David has become an expert catch-and-release fly-fisherman who scoffs at bait fishing, I still enjoy this simpler variety.

The kids and I became extremely enthusiastic fans of college football during the seventies. When watching these games, we would often get so worked up that the neighbors could probably hear us yelling. Although I developed some interest in hockey many years later when I lived in Maine, and I usually watch the World Series, my interest in watching football has declined over the years. Lisa, however, remains an avid fan of various spectator sports, and she also skis. David does not watch many games, but has always participated in surfing and skiing. Aside from this brief period in the seventies, Monique's interests have always been concentrated in other areas.

After we returned from Europe, I was called to Albuquerque to work on the initial design of a huge EMP facility, large enough to test a 747. I enjoyed this activity and put in large amounts of extra time and effort, for which I received some recognition for my work and leadership. My last significant involvement with nuclear effects took place at the atomic test site near Mercury, Nevada. Underground tests were still being conducted at that time and accessing the blast site was an eerie experience. In order to accommodate large equipment, the long tunnel to the blast site was cavernous, with bare, incandescent bulbs lighting the dark earth walls. At the blast site, numerous smaller tunnels radiated outward from the center of the bomb, terminating in walls with direct line of sight of the blast. In order to test the response of small systems, they were mounted on these walls. Companies wishing to test components or systems rented small sections of these walls. I was consulted because the system used to record the response of a component we were testing had failed in a previous test. I went there to implement a design modification that I believed would eliminate the problem; as luck would have it, my design worked.

During most of the next two years, 1975 - 77, I worked in the area of the initial phase of the strategic defense initiative ("star-wars"). In this activity, I led a team consisting of a PhD physicist, a PhD aerodynamicist, and myself. Our assignment was to conceptualize and perform a preliminary performance analysis of all the space and atmospheric ballistic missile designation methods that could be conceived of by ourselves and others. (The purpose of designation is to identify those few of the objects in a ballistic missile attack that contain warheads. An efficient method of fooling a ballistic missile defense or BMD system is to include a large number of lightweight decoys with identical radar and infrared signatures. These decoys would differ from the warhead-containing re-entry vehicle by only their weight and the absence of fissionable material.) I prepared a briefing providing an overview of these numerous (about twenty or so) methods, together with an initial estimation of their viability, and presented it to a group of experts at the Army Ballistic Defense Agency in Huntsville, Alabama. At the last minute, a PhD physicist at McDonnell Douglas other than the extremely competent one on my team had proposed one of the concepts; it appeared to be feasible, and although I did not have the time needed to completely verify it, I included it among the others. The presentation went well until one of the agency's experts noted a flaw in this concept. Discussion of this concept was deferred to a separate meeting, and the remainder of the briefing about all the other concepts went well. (In an odd coincidence, this stroke of bad luck was preceded by an incident the night before, when a hotel staff member awakened me just in time to clear out of my hotel room before it filled with smoke from a fire in a nearby room.) The small crack in my credibility was easily overcome, and notwithstanding the temporary setback, the final report was extremely well received. I personally performed the analyses and wrote the text for nearly half of this report, a large book.

Not long after completion of the final designation report, I was made aware of another concept in this area that had been briefly mentioned elsewhere, but never analyzed or developed further. I extended the concept considerably and developed it to the extent required to demonstrate its viability. I decided to expand this work and present it at a forthcoming national conference. I was aided in this process by one of the individuals who had worked with me on the designation study and two others, one of whom was a master of charts and presentations. Even though I was the principal author and had prepared most of the paper, we both attended the Seventh Annual Department of Defense Conference on Military Laser Technology, which was held at the U.S. Military Academy at West Point, N.Y., where he presented the paper. I answered the questions. This paper is published in the (classified) proceedings of the conference. The paper is entitled: "Threat Information from Range-Doppler Images for BMD Discrimination". The drive from New York City to West Point was a real eye-opener for me. The transition between the bustle of the city and the pastoral beauty of the countryside is surprisingly rapid. I knew the countryside there was beautiful, but I was still a little surprised by it.

We bought a home in Westchester in 1976, retaining our previous home in Mar Vista as a rental. (I was not cut out to be a landlord, and we sold the Mar Vista rental a few years later.) The Westchester house was much larger than our home in Mar Vista, and it was situated on a double lot with a view. The downside was that it needed a great deal of work. After the remodeling experience from Hell, we decorated it with the help of a professional. The final result was beautiful.

The kids spent their teen and college years in Westchester; it was a nice area and they formed many valuable and lasting friendships there. The girls were popular and their boyfriends were well spoken, well mannered, and, at least when I was around, well behaved. David had a couple of long-term girlfriends, but he was also occupied with a love of surfing. I enjoyed these years with the kids; in addition to camping, we took them on a Caribbean cruise as well as to Victoria, British Columbia, via ferry from Port Angeles, Washington, through the San Juan Islands. During this and a subsequent trip to Victoria, we also toured part of the Olympic National Park and Buschart Gardens. (The Caribbean cruise was fun, and I was surprised to find that the water off the Virgin Islands really is the beautiful translucent bluish green of the ads, but it was a little hot and humid for my liking. I enjoyed the ferry trip much more, primarily because we were nearly always near one island or another, so there was more to see in transit.) Within the bounds of normal teenage behavior, they were good and considerate kids. We maintained a strong communication link with them, and thanks to our experience with Danny, they were not able to get away with much.

As a result of my involvement with laser applications, I worked for a short time on aircraft laser and other optical countermeasures. This activity led to the request to present some of my results to personnel at the Naval Weapons Center (formerly known as the Naval Ordnance Test Station) at China Lake, California, which is adjacent to the town of Ridgecrest. On the nearly three- hour drive to this facility, I related the story of the wild teenage party in Ridgecrest and resulting brief group incarceration in the nearby Inyokern jail to the colleague who was accompanying me. After we arrived at the facility and I finished my presentation, I took a rest room break. When I returned, everyone was laughing; to my embarrassment, my companion had told them about my teenage misadventure.

The next important phase of my career began in 1977. I was to be involved in scientific and engineering work in the area of undersea systems for several years, and from time to time thereafter, for a total of about twelve years. This work ranked with the most interesting and exciting of my career. The undersea environment, while intrinsically fascinating, is also particularly interesting and challenging from the viewpoint of sound transmission. Refraction or bending of sound waves results in challenging problems and amazing effects, such as the deep sound channel, through which sound propagation is possible over astounding distances. My primary involvement was with ocean surface and target characterization, sound propagation, ocean noise, sonar signal processing, and sonar arrays. The array work also led me to the study of radar, for which the concepts are mathematically similar. Just as in the case of previous work in signal processing, the primary background required in this work is a good understanding of mathematical statistics and a branch of mathematics known as Fourier transform theory; I had considerable experience in both of these disciplines. (Much of the digital signal processing equations and algorithms that we had to develop at that time are now available in the form of computer chips.) Fourier transform theory is also employed in assessing diffraction effects, one of the fundamental considerations in the study of the propagation of all waveforms, from earthquake waves through sound, radar and light. Whereas refraction deals with the bending of rays along unobscured paths, diffraction theory is used, among other things, to assess the bending of sound waves around obstacles.

During this time period, the company won major study contracts for an advanced lightweight torpedo and a "smart" undersea mine system. Although I did some work on the former, my primary involvement was with the latter. The sonar system subcontractor for the smart mine system was Honeywell Marine Systems Division in Seattle, Washington. As the primary technical interface with Honeywell, I worked with engineers there to develop the sonar system. I was in my element; during a period of about a year, I recall performing and documenting over twenty extensive engineering analyses. The destination of most of my travel at that time was Seattle; I traveled there an average of at least once a month for the next four years. The old waterfront and some of the refurbished areas near the engineering facility retain the essentially maritime character of the city. We were taken to many nice restaurants along the waterfront, and once via ferry to an outstanding gourmet restaurant on one of the tiny San Juan Islands. On yet another occasion, we were treated to a cruise on Lake Washington on the company yacht, complete with open bar and hors d'oeuvres. As part of the engineering work, we also took a small boat out on Puget Sound to conduct measurements of the underwater sound radiated from various ships.

Work on the smart mine system also involved extensive travel to other interesting facilities and cities. Throughout the period we frequently traveled to Washington, D.C., to coordinate the work with Navy factions and gain access to closely-controlled information. For obvious reasons, we usually flew on a McDonnell Douglas DC10. At that time, one carrier had what they called "ocean-to-ocean" service; this service, which was targeted to business travelers, included a brunch bar, giving the flight a somewhat festive feel. When travelling there, we used Washington Dulles International Airport. This meant that you could witness takeoffs of the Concorde supersonic transport; noise considerations dictated that this aircraft had to be flown out of Dulles, which is about 30 miles from central D.C. When sitting on the tarmac with the nose in the drooped position, it looked like some sort of large prehistoric bird. While the sound level of the normal commercial aircraft heard inside the main building is relatively low, when the Concorde fired up, the lower pitched and louder sound was unmistakable, as was the rattling of the huge sections of glass plate forming the walls. When in D.C. we often stayed at or near the Marriot, where beautiful Chesapeake Bay oysters were free and plentiful during happy hour. I soon learned to love raw oysters, and we spent many pleasant, alcohol-laced evenings there, mostly planning or reviewing the day's activities.

I also traveled to the Naval Coastal System center at Panama City, Florida, during this period. This is the primary facility for naval countermeasures. One of the most memorable flights I have ever taken occurred on a DC9 entering this city. Rains in that area on the Gulf of Mexico can be the heaviest I have ever experienced, and this flight was caught in the worst of one of these storms. We were being tossed all over the sky. I was traveling with seasoned veterans of air travel, and the conversational din was never dampened by rough flights, but this was different: the plane was dead silent. Then the plane was hit by lightning. As I found out later, most commercial aircraft are stuck by lightning at least once a year, more in areas of high storm activity. These aircraft are well prepared for this eventuality, which is the result of cloud-to- cloud lightning; the aircraft's skin is designed to conduct this current, and all flight-critical components are well protected. Unfortunately, I did not know this at that time. I was never so happy to get to terra firma. Panama City is a sort of grade B resort town. Oyster shacks were everywhere and the oysters were dirt-cheap; I always had my fill of these. On one of my trips there I was forced to take a return flight some distance away, so I had to take a long drive along the Florida panhandle.

The nearly flat coastal area there has sand in some places nearly as white as snow and many homes are on raised pilings, even though located a considerable distance from the edge of the water. It is particularly interesting to someone accustomed to western coastal areas.

Additional travel during my involvement with the mine program included the Naval Ocean Systems Center in San Diego and the Naval Submarine Base in Groton, Connecticut, where I met with both engineers and a submarine commander, the latter regarding certain submarine tactics. Groton is a typical small coastal town in this region. I almost expected to see an old sea captain on every corner. A severe storm hit the last day that I was there, and when it came time to take my planned flight to Boston in preparation for the trip home, none of the major carriers were yet cleared to fly. I was told that one small carrier was willing to fly, but this evoked the picture of some grizzled old coot with goggles and a scarf in an old biplane. I decided to drive to Boston. This was a good choice. I drove across Rhode Island, stopping in a picturesque little town where I had lunch at a tiny diner that looked like something out of the Forties.

The Navy had funded two competing initial design phases for the smart mine. In the end they liked our competitor's fundamental approach better, and funding for additional study was terminated in early 1981. I worked on a variety of programs over the next nine years, most of which were related to undersea activities. Some of these required a top-secret clearance, which was reactivated from time to time. The first of these activities, however, was considerably removed from undersea work. For the better part of a year, I worked on the problem of communication under the conditions imposed by a nuclear attack. This interesting work involved several trips to a consulting firm in Santa Barbara.

My family held a reunion at my mother's home in Roseville, near Sacramento, in 1982. This was the last time my mom and dad ever saw each other. At about this time I began to have problems with asthma. Since my years of nonsmoking had not allowed me to escape the Pearson family propensity for lung problems, I (irrationatly) decided to smoke an occasional cigarette. I eventually smoked more, and although this was but a small fraction of my former intake, it started to cause me problems, necessitating complete abstinence.

My mother passed away in 1984, my father in 1985. Luckily, I had spent time with both of them prior to the 1982 reunion, as well as visiting with them during this reunion. This was a time of great change in my personal life; I separated from Doris in 1985.

One of the more interesting systems that I worked on in the mid-eighties was a long fiber optic undersea transmission system with multiple repeaters. During this activity, I took a company-sponsored course in fiber optic systems at the Naval Postgraduate School in Monterey, California. In support of later work on a fiber optic network, I traveled to Minneapolis; the weather on my first trip there was perfect. Minneapolis is a pretty city, but evidence of severe winters is apparent everywhere. As a result of the expertise developed while working in the area of fiber optics, I was also asked to present some material on fiber optic networks at the Naval Air Warfare Center in Warminster, Pennsylvania. The countryside around Warminster is dismal in the winter, but the trip afforded me the opportunity to take an enjoyable excursion to nearby Philadelphia.

I worked on various sonar systems during much of the latter part of the eighties. Research funded by Penn State University concerned active sonar, i.e., the design and processing of pulses generated and reflected from targets and other undersea objects, both biological and geographical. In connection with this work, I took two trips to Penn State University, which is located at the town of State College, Pennsylvania. The first trip required a night drive through the snow from Harrisburg. The second trip, in the spring, was via light plane, also from Harrisburg. State College is an upbeat, attractive college town nestled in the green hills of Pennsylvania farm country. Many bars cater to the large student population, and it is pretty active at night. One of the professors invited us to his home for dinner on one of the nights that we were there; he lived in a large old farmhouse in the country some distance outside of town. Of course, we also toured a couple of the bars.

The company was also involved in passive sonar research at that time. This technology concerns surveillance by means of the reception of sound generated by targets, which results from such sources as propellers and engines. Most of the work in this area is concentrated on large undersea arrays. These arrays, coupled with sophisticated signal processing, are capable of extracting a target's sound from great distances in the presence of enormous background noise. Travel during this time included Washington, D.C., San Diego, and Austin, Texas. During part of this period, we were preparing a proposal for a certain type of large array; in this effort we were teamed with a consulting firm in San Diego. My analysis, showing the feasibility of a much smaller array than was proposed by competitors, was one of the key factors in our win. At about this time, I also attended an undersea surveillance seminar at the University of Texas in Austin. One of the highlights of the trip was a real Texas barbecue at the home of one of the seminar participants, near a lake in the nearby countryside.

Near the end of the decade of the eighties, I became involved in another interesting and unusual area of science. My work consisted of the theoretical design and analysis of certain potential applications of an amazing new technological development called Gradar. A well-known member of the guidance community, a PhD employed by Litton, was also involved in the activity. In 1988, we presented our initial findings to various military factions at the Pentagon. One admiral was particularly intrigued by our work, and asked us to travel to Charleston, South Carolina, to present our findings to personnel at the Naval Weapons Station there. We wrote the short classified book: "Gradar Systems and Signal Processing" reporting our findings at the conclusion of the effort. I was the principal author, responsible for most of the work. At the present state of technology, the systems analyzed were prohibitively expensive, but I anticipate that some of this work may eventually be implemented.

I was also involved in a brief study relating to a broad spectrum of phenomena with defense applications in the late eighties, the results of which were presented at company headquarters in St. Louis, Missouri. As part of my involvement in this and other related work during this period, I was chosen to attend a weekend briefing on Soviet technology presented to relevant defense contractors by agents at CIA headquarters in Langley, Virginia.

The divorce terminating my first marriage was granted in January of 1990. By this time, all three of my children held bachelor's degrees. Sometime after the divorce, Doris used some of the proceeds to take the kids on a vacation to the region of her birth in southern Italy.

Second Marriage and Final Career Phase: 1990 - 1998

As stated in the Genealogical Information section, I married Stacey Bailey on May 19, 1990, in Maui, Hawaii. I first met her at work, where I was drawn by her world-beating smile, beautiful eyes and sparkling personality. At that time, I still had most of my typing done by others, and she was one of the few who seemed to enjoy occasionally taking dictation or typing material involving equations. She was attractive, pleasant, competent, had a good sense of humor, and I was fascinated by her descriptions of her place of origin: the faraway land of Maine. I was later to find she had not exaggerated; it really is a wonderful place and culture. During our honeymoon we toured Maui and Kauai.

Notwithstanding the enormous direct and indirect financial loss incurred by the divorce, I had salvaged enough from the settlement to put a down payment on a house. About six months after returning from Hawaii, we bought our present home in the Washington Square area of North Santa Ana. We both like older homes with raised hardwood floors; this one was built in 1938. Like all the homes in Washington Square, this is not a tract house. The place was in pretty sad condition, but it was structurally sound and had an excellent floor plan, including a cellar and a large attic. Luckily, our next-door neighbors are helpful, friendly, and have a general contracting business. With their help, we performed considerable remodeling and refurbishment. Our house was part of a home tour in 1992; it was received with considerable enthusiasm.

During this time period, we often watched our grandchildren Chris and Michelle. The time we spent with them formed some of the happiest of my memories with Stacey. Other memorable times there included great parties and dinners at the Lutz's next door.

Aunt Blanche and Aunt Gertrude both passed away in the early nineties. Fortunately, I had spent time with both of them prior to their death. I was aware of the bad blood between Blanche and my mother, but Blanche liked Stacey and me, so I simply reminded her that if she wished to retain a relationship with us, she would have to keep opinions about my mother to herself. (Considering the tremendous difference between my mother and Blanche, and the failure of my parents' marriage, it is amazing that they ever managed to be civil to one another. Also, Blanche's "in-law" bias likely resulted in her failure to adequately comprehend the role that my father's behavior played in the demise of the marriage.)

The decade of the nineties did not begin well for my career. During the course of an investigation conducted to reinstate my top-secret clearance, it was discovered that I had a couple of security violations resulting from neglecting to lock the file cabinet containing my secret documents. Even though the facility is patrolled with armed guards, this is a reasonably serious offense. The government decided to go after me in a big way, not only refusing to reinstate my top-secret clearance, but also trying to take away all my clearances. A long legal struggle ensued, during which many of my colleagues wrote letters to the government. (A lawyer I briefly consulted about this matter was quite impressed with the letters, joking that a guy would be lucky to receive this much praise at his funeral.) In the end, my top-secret clearance was reinstated.

At this time and through most of the remainder of my career, I worked on electro-optic sighting and designation systems. These systems are able to acquire, track and designate distant targets during both day and night. Target acquisition and tracking is accomplished with arrays of infrared and visible sensors. Designation consists of illuminating targets with a narrow laser beam, thereby allowing them to be destroyed with various laser spot-seeking missiles. Much of this work was very interesting, involving the problems of the propagation of laser, visible, and infrared energy through the atmosphere, control of laser beam characteristics, target detection and tracking and the integration of this capability with aircraft and shipboard systems.

One of the conditions I required before agreeing to work in the area of sighting and designation systems was that the amount of my travel must be greatly reduced, since by this time I was growing a little tired of it. I did, however, add the caveat that I would be quite willing to take a trip or two to England, one of the locations of potential users of this technology. I had to travel only five times over the next eight years, to Portland Oregon, Binghamton New York, Port Hueneme California, St. Louis Missouri and London England. In the portion of the work involving infrared systems, I had meetings with a company in Portland. Presentation of the results of analysis I had performed in support of a proposal necessitated a trip to Binghamton, New York. The Port Hueneme trip was undertaken to provide consultation on integration of the sighting system with shipboard systems. Unrelated work required a trip to headquarters in St. Louis. Target designation work included a trip to London; while I was there, the company we were working with (G.E.C.) took us to the big international air show at Farnborough, England, where we watched the show from their VIP room. (The location of this big international air show alternates with Paris every other year.) The show included performances by the Red Arrows, the RAF's aerobatics team; in a unique twist, commands from the pilot of the lead aircraft were broadcast over loudspeakers. At the time of this London trip, Buckingham Palace was permitting public tours; I took advantage of this opportunity, a memorable experience.

During the target detection analysis work I encountered a problem that had not been adequately addressed in the fundamental literature on the subject. Lamenting the many opportunities to publish that I had not pursued earlier in my career, I decided to explore this. After considerable development and refinement, I submitted the results of this work to the foremost world class journal in this area of technology. The review of the resulting paper, which was conducted by a team led by a professor in Finland, took nearly eight months. (Companies are seldom concerned with the length of the review process for papers of this type, because this has no influence on their utilization of results.) "The paper: Overlapping Search with Scanned Beam Applications", was published in the July 1996 IEEE Transactions on Aerospace and Electronic Systems. (If interested in an abstract, Google: Paul H Duncan ieee scanned beam.)

Although I had some triumphs and bright spots during my work in electro-optic systems, the overall experience was not an entirely pleasant one. I had developed adult-onset asthma in the early eighties, with my first hospitalization taking place in 1983. By the end of the eighties, I had become dependent on Prednisone. This medication, a corticosteroid, produces many undesirable and damaging physical and psychological side effects, such as fat deposits in the face and torso, cataracts, irritability, etc. (This is to be contrasted with anabolic steroids, which are used by athletes to build muscle; the side effects of anabolic steroids include decreased sperm production, blood clots, aggression, etc.) In addition to the hypersensitivity induced by prednisone, the work environment was made somewhat negative by what is best summarized as politics. The deleterious effect on my appearance, reduction in physical strength and compromise in general health resulting from the use of Prednisone also influenced my outlook. All things considered and luckily in possession of enough sense to realize that health is more important than money, I decided to retire. That this option was open to me was in large part the result of Stacey's relatively lucrative position with the FDIC, which hastened a measure of financial recovery from the divorce.

Before I retired, however, I decided to pursue another journal publication and file a patent before I ran out of time. I had not developed this work because I normally devoted much of my "free" time to current work; this practice no longer appeared to be in my best interest. (Although most theoretical work does not lead to patentable results, I had developed but failed to pursue the patent potential of selected work in a few unclassified areas much earlier in my career. Over the years, I had come to regret these missed opportunities.) The desire to complete this work at this time was also motivated by the need to leave a little more of something enduring in the worldwide engineering and scientific society. While I often received satisfaction and recognition during the course of my career, most of this was confined to a comparatively limited engineering and scientific community.

I spent many evenings and weekends completing the paper and patent. The review of the paper took a few months, but it was published well before I retired. The paper: "The Effect of Temporal Image Filters on Observer-Dependent Contrast Metrics" was published in the April 1997 issue of the IEEE (Institute of Electrical and Electronic Engineers) Transactions on Image (digital) Processing. If interested in an abstract, Google: Paul H Duncan ieee temporal. Patent review takes much longer. The patent: "Frequency Measurement Method and Associated Apparatus", U S patent no. 6,026.418, was granted early in my retirement on February 15, 2000. In consonance with the initial design and analysis nature of most of my work, this patent is essentially the implementation of the solution to an interesting but comparatively rare problem. If interested in an abstract, Google: U S patent 6,026,418. A change in emphasis earlier in my career would likely have led to a few more publications; these modest achievements, however, brought some sense of closure to my career. I retired on June 30, 1998, exactly thirty-nine years and six days from the date that I started at the old Douglas Aircraft Missiles and Space Division.

Early Retirement: 1998-2003

I had always planned to retire or change my career to something completely different "sometime", but I derived so much excitement and fulfillment from my work in earlier years that, had this earlier phase continued, I would likely have worked until I dropped. Much earlier, however, I had given the issue some thought from a more philosophical perspective, when I concluded that regardless of how much you may love your work, spending the remainder of your active life involved in a single, all-consuming career constituted a waste of the total life experience. There are, of course, exceptions to the need for retirement or change; if your work is not exhausting or does not require complete involvement, but is still enjoyable, fulfilling and allows extra time to adequately pursue other interests, then there may be little need for major change. There might also be no need for change if your work consists of some world-improving cause in which you are passionately involved. In my case, these exceptions did not apply; all things considered, I am happy that I retired when I did.

Retirement, like a mountain viewed from a distance, appears different when you get there. Like most major life changes, it is an enlightening experience. I knew the adjustment would not be simple but found that it is also an ongoing process. Beyond a few basics, there are no magic formulas for success (a good attitude about aging provides a big advantage, but discussion of this is beyond the scope of this document); the experience is different for each of us.

Unlike previous periods of life during which our parents, school, career, and associated social life provided the primary structure of our activities, when we retire, we are suddenly the only one responsible for defining our activities, and this task continually presents new challenges. Some form of exercise is required to maintain physical health, and it is virtually certain that mental activity retards degradation of mental function. It is also apparent that a certain amount of social interaction is beneficial to mental health. The physical requirements can be met via exercise or recreational athletic activities supplemented with exercise. Assuming a reasonable level of social interaction, mental requirements are met via activities like recreational reading, crafts, solving puzzles, etc. Once the basic requirements are met, the choice of the far greater remainder of our activities is entirely individual. Contrary to some popular retirement concepts, there are no best or most desirable activities, because we are all different. The only requirements are that these activities be actualized rather than contemplated, and in some sense enjoyable to the individual. Although self-motivated activities like mountain climbing, travel, racing, fishing, etc., are obviously enjoyable to most, activities like volunteer work also satisfy the enjoyment criterion via a sense of satisfaction and fulfillment.

The amount of retirement income and savings required in excess of that needed for the basics of life plus some allowance for contingencies depends entirely upon the individual. With the exception of the desire to avoid a significant decline in lifestyle, there are no simple formulas. In my case, a number of personal characteristics and circumstances have mitigated financial requirements. Much of my work provided a great deal of excitement and travel, including the best hotels and restaurants, and I also traveled extensively on my own long before I retired. I also lived my life from the viewpoint that I might lose my health or die at any time, so I did not withhold expenditure with the intent of saving it all in order to make up for missed experiences by lavish spending during the "golden years". Since I had not built up a large life experience deficit when I reached retirement, remedial spending in this area has not been needed. Retirement expense has also been held in check by the fact that, in my case, the increase in free time per se did not lead to a significant increase in expenditure, because much of this time is occupied with activities such as reading. Other personal factors have also reduced the desire for frivolous spending. I have always prized individuality, objectivity and reason; consequently, with the exception of a few fleeting periods during which I failed act particularly logically, I have remained essentially free of the worship of trendy, new, expensive, big and "more" resulting from the envy, greed and trendiness prevalent in American society today. As a result, my happiness is not (above a modest threshold) linked to the quantity or quality of possessions, nor my monetary desires artificially inflated by the need to appear wealthy. (As is well-known but often overlooked, "class", in the commonly understood non academic sense, neither consists of nor is necessarily accompanied by actual or apparent wealth.) After I reached maturity, the only financial limitation that ever significantly troubled me was the inability to supply greater financial assistance than that which I was able to provide for my children at a couple of critical points in their lives. Unfortunately, these times coincided with temporary but extended periods of financial problems.

Just as during other phases of life, my retirement interests change frequently; the only potentially expensive interests that seem to be lasting longer than usual are occasional foreign travel, some additional extensive domestic travel, and possibly a newer motor home. I am entertaining the possibility of buying a newer RV, provided I can find one a little smaller than my present (23-ft) one, and with nearly as much window space for viewing the scenery over morning coffee. (I hate maneuvering large vehicles in the unavoidable trip segments through the cities, and larger motor homes are not allowed in many of the smaller, pristine campgrounds. If not for the fact that this size range has just sufficient space for comfortable indoor socializing in inclement weather, I would prefer a much smaller one.) All things considered, my Boeing pension augmented by Social Security and modest contingency savings allows us to maintain our comfortable pre-retirement lifestyle and do most of the things that interest us.

Stacey and I had long discussed retiring in Maine. Sadly, Stacey's mother passed away before we were able to retire, but her father still lived there and she had some old friends and other family there. We decided to rent our home to her daughter, Bonnie, and move to South Portland, Maine. Our home in South Portland was located on a tree-lined street a little over half a mile from the ocean. A large wooded area adjoined the back of the property, and it was only a few minutes to the country.

Spring and fall on the southern coast of Maine are beautiful. A few days in summer are excessively warm, but air conditioning is not needed. Winter is not nearly as cold on the coast as it is inland, but the temperature still typically drops to somewhat below zero overnight more than once before the arrival of spring. We used to love watching it snow from the warm comfort of our home and walking in the woods after a snowfall. Although it is cold, storms move quickly and at least half of the winter days are sunny.

During the first (1998) fall in Maine, I took a brief trip to the West Coast to attend the wedding reception for my youngest daughter Monique and her husband Matt (Mathew) Gould. After their marriage in Italy and a honeymoon tour that included the U. K. and Ireland, this event was held in their home city of San Francisco. They both like to travel. Monique had studied in Italy for six months, taking some of this time to tour Europe, and later spent time in Bali during several trips associated with her work. She and Matt recently took an extensive trip to Russia, India, and Tibet. Matt had connections in Russia and India through colleagues in his work. While in Russia, they were escorted by one of these friends, and attended the Bolshoi Ballet; during their stay in India, they stayed with the friend of a colleague. Monique is involved in various art-related activities, and Matt has completed his internship and examinations and has set up a practice as a licensed psychotherapist. They recently (late in 2002) bought a large apartment, consisting of the ground floor of one of the older buildings in San Francisco, which they are busily refurbishing and redecorating.

It is fortunate that we rented our home in Santa Ana, because midway during our first winter in Maine, Stacey informed me that we would have to go back. She had not realized how much her mother's death, the changes brought by years of separation from

Maine, and her separation from her daughter (who lived minutes away when we were in Santa Ana) had altered her perception of retirement there. This considerably changed my approach to our time in Maine. I had intended to fly to the West Coast fairly frequently to visit my family, but now we had to save, not knowing how much of a loss we might incur on the sale of the house. As it turned out, when we sold two years later, we actually came out slightly ahead financially.

Early in our first (1999) summer in our home in Maine, I traveled to Los Angeles to attend my oldest daughter Lisa's master's degree graduation ceremony. As part of this trip I also went to Orange County to see Stacey's daughter Bonnie and her family, as well as our good friends next door. On the way home I stopped in Salt Lake City to see my son David and his fiancée Linda. While I was there, David and I went fly-fishing and rafting on the Green River with one of his friends. We had a minor mishap on the river when a submerged tree tore a large hole in the raft, but we escaped uninjured. (This river has not been good for me. As a result of an accident during a previous fishing trip there, the pupil of one eye is permanently dilated. I also had to replace the lens in that eye, but this was likely inevitable anyway, because I had developed some cataract there as a result of long-term Prednisone use.) When I returned to Maine, the woods adjoining our back yard were replete with wild blueberries, and Stacey and I spent many pleasant afternoons picking them.

One of the options I considered when I retired was part-time consulting. Given the right opportunity during the first two or three years of retirement, I would probably have enjoyed it. No such opportunity presented itself in Maine, and I had no desire for distant assignments. Some people at Boeing called me several times with various questions, and during our second fall in Maine, they contacted me about a consulting assignment to begin the following year. This work never materialized, evidently because the bureaucrats failed to approve my requirements that most of the work be performed at home and require no more than two or three months of my time. By the time I returned from Maine, I had decided that I no longer had any desire to add this kind of work-related stress to my life, so when a former colleague indicated the interest of another company, I declined. At the time of writing, I also have no desire to enter any commitment requiring the relinquishment of a significant amount of my freedom, whether financially compensated or not. As stated earlier, however, retirement is not static, so I may yet enter into some kind of limited commitment.

We spruced up the Maine house quite a bit and I built a large deck adjoining the back of the house, completing it before our second winter there. The deck was finished in time for a visit from David and Linda in the fall of 1999. After a brief local tour I loaded Linda, Stacey, David, and our dog and two cats into the RV and took a scenic trip to Nova Scotia. Although Linda had to return to her job in Salt Lake at the conclusion of this trip, David was able to remain long enough to install a door leading from the kitchen to the deck. Later, we had visits from Elsie and Maurice, some friends from Ohio, Lisa and Todd, and my brothers, David and Ray.

After Linda and David took a vacation trip together to Costa Rica, and David had a final bachelor fling in Australia and New Zealand, they married in the spring of 2000 and moved into her nice, attractively decorated home in Salt Lake City. I flew out to the wedding from Maine. The cleric conducting the service gave the best talk on marriage that I have ever heard. The location chosen for the ceremony was beautiful, and the ceremony was well attended and unique. They both love animals, and they had their two dogs released to them at the proper moment during the ceremony, with rings contained in purses attached to their collars. Linda is completing her BA; she plans to work in the medical field and is considering studying for a master's degree to become a nurse-practitioner. David has run his own construction business since 2001, and they recently purchased another house for speculation and rental income. He makes twice as much as when he worked for others and has more time for fishing, skiing and his hobby of restoring old trucks. The time he spends at these activities, however, will soon become more limited: I recently (early Summer of 2003) learned that Linda is pregnant with my first direct descendant grandchild, due when she completes her degree in January of 2004.

The sale of our Maine house took a little longer than expected, so we were not able to leave for California until October of 2000. The trip included visiting friends in Ohio, Linda and David in Salt Lake City, and a side trip to see friends in Idaho.

When we returned to our home in Santa Ana, we found that Bonnie was experiencing considerable difficulty in finding a new home; they needed five bedrooms for their large family. She finally found a large house near our home, and the family moved out in June of 2001. Stacey and I see Bonnie and her family often, sometimes baby-sitting Max. Although these children are officially my step-grandchildren, we have always been very close and I have always thought of them as simply my grandchildren. Bonnie and her husband, Mike, have a successful plumbing contracting business. In addition to running this business, Bonnie is deeply involved with all of the children's activities, including Michelle's cheerleading -- Michelle is now a coach for the award-winning Fountain Valley cheerleading team. Mike, who calculates the job bids and supervises as well as performs much of the plumbing work, is also heavily involved in coaching Mikey's baseball and football teams.

During the time that Bonnie was looking for a house, Stacey and I stayed in our upstairs den. We did a little traveling and I took the opportunity to read a lot. In early May of 2001, we flew to Maine to pick up a car that Stacey's dad bought for us and spent the rest of the month in a leisurely and enjoyable drive back. The route took us by Niagara Falls, the home of friends in Ohio, and Chicago, after which we drove on or paralleled Route 66, with side trips to Taos and Sedona. We also took the opportunity to visit the Painted Desert and Grand Canyon, and spent a few days in Las Vegas, where Stacey won more than enough to cover our expenses in that city.

Bonnie completed her move when we returned to Santa Ana, and we were able to let our cats out for the first time. (They had a pit bull, so we were not able to let them out during the entire time that we were cloistered in our upstairs den.) I have always liked animals, but was always too busy to pay much attention to them; since retiring, however, my affection for these fellow travelers in life's brief journey has added a new and pleasant dimension to existence. One week after letting our cats out, my all-time favorite pet, a very special little cat who used to walk with our golden retriever and me in the woods behind our home in Maine, disappeared. We had weathered the painful loss of other pets recently, but her disappearance prevented closure; it took me months to recover from the loss. Pets always evoke a parental response, but she evidently brought more of that out in me. More significantly, she provided a special little ray of sunshine during the difficult period of adjustment to retirement; I had come to feel, however irrationally, that the happiness she brought was the little extra spark I needed to get through this period.

We had a somewhat limited Pearson descendants' get-together at Sequoia in August of 2001. A second, more complete family reunion was held in Ahwahnee, California, in August of 2002. I met my niece, Lacey, for the first time there (I saw her at the '82 reunion, but she was a baby). A golden (50th) wedding anniversary celebration for Elsie and Maurice was hosted by their sons in Las Vegas in January of 2003; most of those who had been at the 2002 reunion were there, plus many more friends, some dating back to high school. I also met my nephew, Ray, for the first time at this event.

In December of 2001, Lisa married Todd Daly in a beautiful and well-attended ceremony at a ranch in Malibu. They had obtained their master's degrees at the same time, and they both work in psychology-related fields. Lisa has traveled quite a bit, including New Zealand, the Netherlands, Greece, France, Italy and Spain, and she also lived for a year in Aspen, Colorado. Todd's travel experiences are more limited, but they intend to rectify this omission. Prior to the present (August of 2003), they have lived in an apartment in Brentwood; it has been nice to be able to visit them more frequently and go RV-camping with them. Considerations of professional opportunity have now dictated that they move to Palo Alto; it is hoped that we will still be able to visit them at about the same frequency.

With the aid of the Internet, encyclopedia and books borrowed from a large nearby library, I enjoy studying virtually everything and reading the American, British and Continental classics that I have missed. Stacey and I also go camping, work on the house, travel, and spend time with relatives and friends. In addition to fishing now and then, I have initiated several other activities of a hobby nature, but none of these has led to significant involvement; Stacey is usually involved with her shopping, antiquing and decorating hobbies.

In May-June of 2002, we took a marvelous month-long tour of Paris, Versailles and the UK. On the advice of a Francophile friend who had spent a decade in Paris, we stayed in the Latin Quarter near Notre Dame, about a block off the Seine. After nearly a week in Paris, we took the "chunnel" train beneath the English Channel to London, where we rented a car for a leisurely tour of England and Scotland, traveling from one B & B to the next. In March of 2003 we took an interesting guided tour (my first, other than one-day tours) of China that included Beijing, Shanghai, Suzhou and Hangzhou. Although I prefer travelling on my own and learning a few phrases, mingling with locals, and generally experiencing more of the life and culture of a region, the difficulty of the Chinese language rendered this more difficult than I am prepared to undertake. (Although it is possible to find people who speak English almost everywhere, it is not always easy.) We were lodged at large, plush hotels throughout the tour. I have always enjoyed B & Bs and small hotels with a lot of character, regarding big hotels as appropriate for stay-in vacations or work and event convenience, but they are also very desirable in countries like China, with poor or questionable sanitation and food handling practices.

All things considered, we are thoroughly enjoying retirement. I try to live in the present and enjoy each day, include as much variety in my activities as I can comfortably handle, and make no plans that cannot be changed. I am also avoiding commitments that significantly constrain my freedom, but this may change with time.

This is the end of the life story written in 2003. The following is a brief overview of events since then.

Developments in Descendants' Lives: 2003 to 2020

As detailed in the genealogy section, a number of grandchildren and great-grandchildren of Laura Pearson's children were born after 2003. Some (but far from all) of the other life changes, milestones or interesting events since that time are:

My wife, Stacey passed away and I live in Palm Springs; Doris Duncan lives in Mountain View, California; Maurice and Elsie installed all the electrical and plumbing in the Red Mt. property needed to support large gatherings on the large property, which the Kennedy family retains; Elsie's husband Maurice passed away and she now lives in Arroyo Grande, California; Ray Rogers completed a master's degree in 2003; he and wife Akemi travel a lot and he volunteers for the Red Cross; Mark Rogers has retired and is doting on his new granddaughter; Michelle Hanson, now married to Jason Rothgeb, has just completed a bachelor's degree; Bonnie Hanson contributed a chapter to the book "Inspiration for Writers"; Chris Hanson started a new business; Lisa (Duncan) Daly and husband Todd are now practicing psychotherapists living in Menlo Park, California; Greg Rogers achieved Eagle Scout status and went on to complete medical school and flight school in the U S Navy; he will begin his residency next year; Bret Davis completed a master's degree; my son David and his wife Linda have divorced but are both doing very well; Several of Monique Duncan-Gould's floral creations-to-art have been shown at San Francisco museums; Lacey (Duncan) Rice, a surgical nurse, worked at hospitals in California and Michigan before joining the pediatric open heart surgery team at the University of Kentucky; she married Michael Rice in 2018 and now lives in Las Vegas; Lee Kennedy is selling his business and performs expert witness depositions; he spends much of his time at a second home in Arizona; Danny Kennedy is living in Red Mountain; Todd Kennedy has retired and is living in Nevada; Douglas Kennedy is now married; Tori Kennedy is now married to Jason Baker; Brianna Kennedy is now married to Alex Romero; Allison Kennedy obtained a bachelor's degree in 2007 and is now married to Shaun MacDonald. Sadly, in addition to the aforementioned deaths of Stacey Duncan and Maurice Kennedy; Ray Duncan, Diane Duncan Weston, Melva (Andrew) Norris and Tatum Kennedy have also passed away.

Retirement, Continued: 2003-2020

We sold our Santa Ana home in 2004 and used the proceeds to buy places in Windham, Maine and Palm Springs, California. We spent summers in Maine and winters in California. The home in Windham is located on the shore of a beautiful pond and includes a dock. The place in Palm Springs is a new and up to date modular home with three bedrooms and two bathrooms. It is located in a manufactured home park. Winters there are idyllic.

The home in Windham needed a lot of work to make it livable, but we loved spending summers there. A close friend of Stacey's from high school lives nearby with her husband and we spent a lot of time together. We also visited with other friends from Portland, and Stacey held a reunion with her Pi Kappa high school sorority at the Windham home. We also enjoyed spending winters in Palm Springs, and soon made friends there.

The financial and other problems with the extensive work on the Windham house, the travel with four animals twice a year and the double expenses for cars, insurance etc. began to take a toll. We decided to sell that home. While awaiting the sale, Stacey became ill and required an operation including the installation of a colostomy bag. We sold the Maine place in 2010 and settled into living full time in Palm Springs. Soon after arriving there, Stacey had an operation completing the previous surgery and removing the need for the colostomy bag.

Our marriage started to have problems during the last few years in Maine, and when we settled in Palm Springs, we began to drift further apart. I approached her about seeking counselling, but she did not believe in it. I naively thought things would somehow get better, so I was not prepared for the shock of learning that she wanted to leave. When I finally came to grips with this, I did not attempt to stop her. I offered her as much financial aid to augment her social security and small pension as I could, but she was not able to find a suitable place to stay. Complicating and contributing to all this, her symptoms of bipolar disorder and other more serious issues worsened, as did my age-related and other health problems. I adopted the position of a caregiving family member until a more permanent solution could be established.

While attending a life celebration for my brother in law Maurice in the spring of 2017, I suffered a spell of dizziness and became somewhat short of breath. It was subsequently determined that I had a form of atrial fibrillation that causes dangerously extended periods (20 minutes or more) of tachycardia, or pulse rate exceeding two hundred beats per minute. This required placing me on medications to slow the heart. To prevent the excessive slowing that this medication would cause, it was necessary to install a pacemaker. Complications related to the anticoagulant I also had to take, and some as-yet undetermined by-product of this treatment resulted in a slow internal bleed.

One of the symptoms that Stacey exhibited was what appeared to be TIAs (transient ischemic attacks or temporary strokes). On the last incident of this type, her mouth and voice became stuck for a short time, mid-syllable. After some discussion, she decided to get an examination to diagnose the problem. After hours of waiting in the emergency room, she was finally admitted to a room. Following extensive examination and questioning, it was determined that the blood supply to her brain was likely inadequate. The recommended treatment was the implantation of a stent in her aorta. The doctors said that the probability of problems was about three percent. We pretty much assumed they meant either a stroke or no improvement in symptoms, but we weren't too worried because she was on the anticoagulant aspirin. (Patients undergoing this procedure are required to take an anticoagulant.) Stacey had undergone many operations before and wasn't worried. I actually suggested that one option was to forego the procedure, but she immediately vetoed that, and the nurse with her concurred. Days of waiting followed, including examinations, questions and delays to get the right size tube for the stent. After one final x ray, she was wheeled back to her room before the operation. As her bed rounded the corner and she saw me, her face lit up with that indescribably world-brightening smile, the last I ever saw. On the way out of the hospital, I stopped at the gift shop and ordered some flowers for her. She was not worried and had successfully convinced me to not worry.

I called the hospital the next day at about the time in the afternoon when she would be expected to awake from the anesthetic. The nurse cheerfully told her who was calling and handed her the phone. Stacey made a little sound as though still groggy but did not talk, so I decided to call later. I was not alarmed but did not feel well, so I thought I would rest and go see her in the morning.

The first call from the hospital came late in the evening of the same day. The doctor asked my permission to install a drain port to relieve pressure in the skull. I gave them the okay. Evidently when the stent suddenly allowed increased blood flow, an artery in the brain that had weakened over time must have ruptured, causing a cerebral hemorrhage. I had no more recovered from this news and finished dressing to go there when the second call came. They asked me to come in because "the patient's condition is deteriorating". I rushed there, only to find that she had become brain dead. After some time, I gave permission to withdraw life support, and held her hand as she passed on the early morning of February 2, 2018, at age 73. Bonnie arrived with Michelle a little while later. Shortly after that I kissed Stacey goodbye and we left. I was so distraught that I was unable to drive, so Michelle drove me home in my car.

There are no words that adequately describe the shock, horror and grief at the sudden, completely unexpected and untimely death of a loved one with whom you have spent more than thirty years. The fact that we had not gotten along well for some time did not diminish this. I only felt as though I wanted to die.

The next morning it was decided to have Stacey's remains cremated (as I plan for myself) and hold a life celebration later. (The celebration, which was held a few months later, was quite beautiful and tasteful.)

My physical condition rapidly deteriorated and it was decided to perform a colonoscopy to find the slow bleed. Likely because I had not eaten and my hemoglobin was low, when I went in for the colonoscopy, I passed out cold in the waiting room. Bonnie, who had accompanied me, evidently partly cushioned the fall. When I was under anesthesia, the doctor also performed an endoscopy. Both procedures were negative for blood. Afterward I was unable to stand and was kept in the hospital for more than a week. Lisa, Ray and Elsie helped me while I was in the hospital. I later underwent a procedure to examine the small intestine for blood. This was done by requiring me to swallow a small camera. This procedure was also negative for blood. Bonnie was with me for all these procedures and for some time afterward while recovering at home. After that, Monique was able to stay with me for a while.

The following year was darkened by bleak episodes of grief. During this time, I separated about a third of Stacey's ashes for shipment to her sister. I had the container engraved with the words: "Our beloved ray of sunshine". Thanks to extended family visits, assistance and support, plus a daily visit here from my best friend and neighbor Michael Meyer, I gradually improved. About 15 months after Stacey's death, I attended a big Kennedy-Rogers-Duncan Easter get-together at the Kennedy compound in Red Mountain and, seven months later, enjoyed Thanksgiving dinner with Lisa and family at my place in Palm Springs. I capped 2019 by spending Christmas at Bonnie's. I had begun to feel almost normal by that time when I had to have my favorite dog euthanized. By the spring of 2020, I was almost ready to send Stacey's sister the ashes that I had prepared for her, then along came Covid-19.

Alone at home, self-quarantining and awaiting a vaccine since March 12, 2020, I am grateful that I feel reasonably well on most days. The sight in my long-ago injured left eye has finally failed completely, but I can still drive and otherwise function satisfactorily. Bonnie braved hours of stalled traffic to bring home-cooked food, gifts, cupcakes and, last but not least, our little granddaughter, Dallas, to help celebrate my 85th birthday on August 27. All three of us wore masks throughout the visit and sat about seven feet apart; a lot of trouble, but with my heart and lung problems and age, it is necessary.

In writing this story, I have emphasized positive or humorous events, times, accomplishments or stumbles in order to attempt to make it more interesting. I really do not think anyone wants to read about all my mistakes, problems, faults, physical and mental health issues, etc. The intent was only to share with you some of the experiences that I judged to be interesting or otherwise worth noting.

Reviewing my life as I write this, I have mostly pleasant memories of my life and especially my parents, wives, children, siblings, grandchildren, great-grandchildren, the rest of my extended family, and those I have had the good fortune to have as friends.

Made in the USA
Monee, IL
04 November 2026

67384592R00069